Advanced Energy Design Guide for Small to Medium Office Buildings

This publication was prepared under the auspices of ASHRAE Special Project 133.

PROJECT COMMITTEE

Bing Liu
Chair
Pacific Northwest National Laboratory

Merle McBride
Vice-Chair
Owens Corning

Don Colliver
Steering Committee Ex Officio
University of Kentucky

Daniel Nall
AIA/USGBC Representative
WSP Flack + Kurtz

Erin McConahey
Member at Large
Arup

Mick Schwedler
ASHRAE Representative
Trane, a division of Ingersoll Rand

Brian Thornton
Analysis Support
Pacific Northwest National Laboratory

Michael Lane
IES Representative
Lighting Design Lab

Weimin Wang
Analysis Support
Pacific Northwest National Laboratory

Mathew Tanteri
IES Representative
Tanteri + Associates

Lilas Pratt
Staff Liaison
ASHRAE

STEERING COMMITTEE

Don Colliver
Chair

Bill Worthen
AIA Representative

Jerome Lam
DOE Representative

Rita Harrold
IES Representative

Mick Schwedler
ASHRAE SSPC 90.1 Liaison

Brendan Owens
USGBC Representative

Adrienne Thomle
ASHRAE TC 7.6 Liaison

Tom Watson
ASHRAE Representative

Lilas Pratt
ASHRAE Staff Liaison

Advanced Energy Design Guide for Small to Medium Office Buildings

Achieving 50% Energy Savings Toward a Net Zero Energy Building

American Society of Heating, Refrigerating and Air-Conditioning Engineers
The American Institute of Architects
Illuminating Engineering Society of North America
U.S. Green Building Council
U.S. Department of Energy

ISBN 978-1-936504-05-3

Library of Congress Cataloging-in-Publication Data

Advanced energy design guide for small to medium office buildings : achieving 50% energy savings toward a net zero energy building / American Society of Heating, Refrigerating and Air-Conditioning Engineers ... [et al.].
    p. cm.
  Includes bibliographical references.
  Summary: "Designed to provide recommendations for achieving 50% energy savings over the minimum code requirements of ANSI/ASHRAE/IESNA Standard 90.1-2004 for small to medium office buildings; allows contractors, consulting engineers, architects, and designers to easily achieve advanced levels of energy savings without having to resort to detailed calculations or analyses"--Provided by publisher.
  ISBN 978-1-936504-05-3 (softcover : alk. paper)
  1. Office buildings--Energy conservation. 2. Office buildings--Design and construction. I. American Society of Heating, Refrigerating and Air-Conditioning Engineers.
  TJ163.5.O35A38 2011
  725'.0472--dc22

                                                    2011006680

ASHRAE STAFF

SPECIAL PUBLICATIONS

Mark Owen
*Editor/Group Manager
of Handbook and Special Publications*

Cindy Sheffield Michaels
Managing Editor

James Madison Walker
Associate Editor

Elisabeth Parrish
Assistant Editor

Michshell Phillips
Editorial Coordinator

PUBLISHING SERVICES

David Soltis
*Group Manager of Publishing Services
and Electronic Communications*

Jayne Jackson
Publication Traffic Administrator

PUBLISHER

W. Stephen Comstock

Contents

Sidebars—
Case Studies and
Technical Examples

Acknowledgments

Advanced Energy Design Guide for Small to Medium Office Buildings is the first in a series of Advanced Energy Design Guide (AEDG) publications designed to provide design strategies and recommendations to achieve 50% energy savings over the minimum code requirements of ANSI/ASHRAE/IESNA Standard 90.1-2004, *Energy Standard for Buildings Except Low-Rise Residential Buildings*. This 50% AEDG series addresses building types that represent major energy users in the commercial building stock. The publication of this Guide is the result of collective efforts of many professionals who dedicatedly devoted countless hours in writing this book.

The primary authors of this Guide were the nine members of the ASHRAE Special Project 133 Committee (SP-133). The chair would like to personally thank all the members of the project committee, who worked extremely hard to pull together practical, technically sound information covering the full range of integrated designs of ultralow-energy-use office buildings. The authors brought to this project many years of experience and good practices in design, construction, commissioning, and research of office buildings to achieve significant energy savings.

The project committee met four times and held 12 conference calls in six months. Most of the face-to-face meetings were scheduled on weekends so the committee members could still meet deliverables and expectations from their daily paid jobs. Thus, the chair would also like to express her appreciations to the authors' families for their sacrifice and support of this project. In addition, I'd like to acknowledge the support from the employers of the project committee members, including Owens Corning, WSP Flack + Kurtz, ARUP, Lighting Design Lab, Trane Company, Tanteri + Associates, the University of Kentucky, and Pacific Northwest National Laboratory (PNNL). The project would not have been possible without the financial contributions of the U.S. Department of Energy (DOE), through the Building Technologies Program.

The project committee's efforts were guided by the AEDG Steering Committee (SC), made up of members from the partner organizations: American Society of Heating, Refrigerating and Air-Conditioning Engineers (ASHRAE), American Institute of Architects (AIA), Illuminating Engineering Society of North America (IES), U.S. Green Building Council (USGBC), and DOE. The SC assembled an expert team of authors and defined a project scope that kept the project committee's task manageable and focused. The representatives from these organizations brought a collegial and constructive spirit to the task of setting policy for the Guide.

The AEDG SC convened a focus group of building owners, investors, architects, designers, contractors, and maintenance staff to help guide the overall concept of the document (i.e., the scoping document). The focus group members who provided valuable insight into the needs of this Guide are Mark Pojar of Hines Conceptual Construction Group, Rives Taylor of Gensler, Brenda Mowara of BVM Engineering, Steve Winkel of the Preview Group, David Okada of Stantec, Zeina Grinnell of Beacon Capital Partners, Clark Manus of Heller Manus Architects, Morgan Gabler of Gabler-Youngston Architectural Lighting Design, Chris Green of Ago Studios, Tim McGinn of Cohas-Evamy, and Brian Livingston of Haselden Construction.

In addition to the voting members on the project committee, there were a number of other individuals who played key roles in the success of this Guide. The specific individuals and their contributions are: Brian Thornton and Dr. Weimin Wang of PNNL for writing contributions to Chapters 3 through 5 and extensive building simulation analyses and results; Lilas Pratt, Mick Schwedler, and Michael Lane for serving as the gracious hosts of the meetings at their facilities; and Bill Worthen of AIA for his focus on Integrated Project Delivery and direct contributions to Chapter 2. This Guide could not have been developed without all of their contributions.

Twenty people participated in the two peer reviews, providing more than 360 remarks that helped to strengthen and clarify the Guide. We appreciate the considerable time they took from their busy schedules to give us their thoughtful input and hope that they see the impacts of their recommendations in the finished product.

A huge debt of gratitude is extended to the authors of the previously published 30% AEDGs because they paved the way and defined the basic structure, content, and format of the books as well as the procedures for the reporting and the reviews. Following in their footsteps has provided consistency among the AEDGs in addition to being a tremendous time saver. Building upon the previous work also allowed the project committee to finish its work in a very short period of time.

Lilas Pratt, ASHRAE Special Projects Manager, went beyond the call of duty in managing an enormous number of documents and coordinating with all the authors with great competence and efficiency. Cindy Michaels, Managing Editor of ASHRAE Special Publications, did a tremendous job in making the document into a first-class publication. Their efforts, as well as those of many other ASHRAE staff persons, are greatly appreciated.

I am very proud of the Guide that the project committee developed and amazed at the accomplishment in such a short time period. I hope you find this Guide provides resourceful and practical information in guiding your energy-efficient office building design.

Bing Liu
Chair, Special Project 133

January 2011

Abbreviations and Acronyms

100% OAS	100% outdoor air system (also *dedicated outdoor air system* [*DOAS*])
AEDG-SMO	*Advanced Energy Design Guide for Small to Medium Office Buildings*
AHRI	Air-Conditioning, Heating, and Refrigeration Institute
AIA	American Institute of Architects
ANSI	American National Standards Institute
ASHRAE	American Society of Heating, Refrigerating and Air-Conditioning Engineers
ASTM	ASTM International
BAS	building automation system
BAU	business as usual
BEF	ballast efficacy factor
BEF-P	ballast efficacy factor—prime
BF	ballast factor
BoD	Basis of Design
Btu	British thermal unit
C	thermal conductance, $Btu/h \cdot ft^2 \cdot °F$
CBECS	Commercial Buildings Energy Consumption Survey
CD	construction document
CDD	cooling degree-days
cfm	cubic feet per minute
CHW	chilled water
c.i.	continuous insulation
CO_2	carbon dioxide
COP	coefficient of performance, dimensionless
CRI	Color Rendering Index
CRRC	Cool Roof Rating Council
Cx	commissioning
CxA	commissioning authority
DCV	demand-controlled ventilation
DL	Advanced Energy Design Guide code for "daylighting"
DOAS	dedicated outdoor air system (also *100% outdoor air system* [*100% OAS*])
DOE	U.S. Department of Energy

DX	direct expansion
EA	effective aperture
ECM	energy conservation measure
EER	energy efficiency ratio, Btu/W·h
EF	energy factor
EIA	U.S. Energy Information Administration
EL	Advanced Energy Design Guide code for "electric lighting"
EN	Advanced Energy Design Guide code for "envelope"
ERV	energy recovery ventilator
ESP	external static pressure
E_t	thermal efficiency, dimensionless
EUI	energy use intensity
F	slab edge heat loss coefficient per foot of perimeter, Btu/h·ft·°F
fc	footcandle
FC	filled cavity
FFR	fenestration to floor area ratio
GSHP	ground-source heat pump
Guide	*Advanced Energy Design Guide for Small to Medium Office Buildings*
HC	heat capacity, Btu/(ft^2·°F)
HDD	heating degree-days
HID	high-intensity discharge
HSPF	heating seasonal performance factor, Btu/W·h
HV	Advanced Energy Design Guide code for "HVAC systems and equipment"
HVAC	heating, ventilating, and air-conditioning
IAQ	indoor air quality
ICC	International Code Council
IEER	integrated energy efficiency ratio, Btu/W·h
IES/IESNA	Illuminating Engineering Society of North America
IGU	insulated glazing unit
in.	inch
in. w.c.	inches of water column
IPD	Integrated Project Delivery
IPLV	integrated part-load value, dimensionless
kW	kilowatt
kWh	kilowatt-hour
LBNL	Lawrence Berkeley National Laboratory
LCCA	life-cycle cost analysis
LCD	liquid crystal display
LED	light-emitting diode
LEED	Leadership in Energy and Environmental Design
LPD	lighting power density, W/ft^2
LPW	lumens per watt
Ls	liner system
LZ	lighting zone
MERV	Minimum Efficiency Reporting Value
N/A	not applicable
NEMA	National Electrical Manufacturers Association
NFRC	National Fenestration Rating Council
NV	Advanced Energy Design Guide code for "natural ventilation"
NREL	National Renewable Energy Laboratory
OA	outdoor air
O&M	operation and maintenance
OPR	Owner's Project Requirements

PF	projection factor, dimensionless
PL	Advanced Energy Design Guide code for "plug loads"
PNNL	Pacific Northwest National Laboratory
ppm	parts per million
PV	photovoltaic
QA	quality assurance; Advanced Energy Design Guide code for "quality assurance"
R	thermal resistance, $h \cdot ft^2 \cdot °F/Btu$
RE	Advanced Energy Design Guide code for "renewable energy"
rh	relative humidity
ROI	return on investment
SAT	supply air temperature
SEER	seasonal energy efficiency ratio, $Btu/W \cdot h$
SHGC	solar heat gain coefficient, dimensionless
SRI	Solar Reflectance Index, dimensionless
SWH	service water heating
TAB	test and balance
U	thermal transmittance, $Btu/h \cdot ft^2 \cdot °F$
USGBC	U.S. Green Building Council
VAV	variable air volume
VFD	variable-frequency drive
VRV	variable refrigerant volume
VSD	variable-speed drive
VT	visible transmittance
W	watt
WH	Advanced Energy Design Guide code for "service water heating"
WSHP	water-source heat pump
WWR	window-to-wall ratio

Foreword:
A Message for Building
Owners and Developers

Advanced Energy Design Guide for Small to Medium Office Buildings is the first in a series of Advanced Energy Design Guide (AEDG) publications designed to provide recommendations to achieve 50% energy savings when compared with the minimum code requirements of ANSI/ASHRAE/IESNA Standard 90.1-2004, *Energy Standard for Buildings Except Low-Rise Residential Buildings* (ASHRAE 2004). This guide applies to small to medium size office buildings with gross floor areas up to 100,000 ft^2.

Use of this Guide can help in the design of new office buildings and major renovations that consume substantially less energy compared to the minimum code-compliant design, resulting in lower operating costs. Also important, through use of an integrated design process, an energy-efficient building offers a great possibility to enhance the working environment, including indoor air quality, thermal comfort, and natural lighting. Studies show an enhanced working environment can increase worker productivity, improve staff retention, increase corporate brand awareness, have significant rental and sales prices premiums, and may contribute directly to corporate social responsibility through reduced greenhouse gas emissions (RMI 1994; Costar 2008).

The ability for a project to achieve and maintain 50% energy use reduction in any climate and on any site requires more effort than project design team agreement on any specific set of building energy systems. Fundamental building program requirements by the owner need to be considered prior to making even basic architectural design decisions. Considerations include the nature of project team contracts; a basic understanding of climate, site requirements, drivers for building orientation, and massing; and project budget/cost assumptions of the building owner and/or their designated construction manager. The project team should also set a goal to allow for ongoing maintenance and replacement of critical systems to ensure the enduring performance of the overall building design.

One of the goals of this Guide is to provide useful information about energy-efficient system design and how the performance of these systems relates to decisions made by the owner, from site selection, initial project budget, request for proposal language, and Owner's Project Requirements to many of the building design decisions made very early in concept designs. There is also a direct relation of the skills and understanding of individual design team members as well as their awareness of concepts included in this guide and their ability to work collaboratively, share risk, and have sufficient time and fees to achieve 50% energy use reduction. It is also worth noting here that whatever time and effort is spent by the project's design team to achieve this level of efficiency has a high risk of being negated or lost if the building's operation and maintenance staff is not provided with the tools, education, information, and decisions made during the initial design process, including the assumptions made concerning equipment, maintenance, calibration, and replacement of critical building systems. Ultimately, the design team has no control of the "life" and "use" of the building by its occupants beyond the scope of their contract for services.

This Guide presents a broad range of subject matter, including broad concepts such as the integrated design process, multidisciplinary design strategies, and design tips and good practices on specific energy systems, while the focus of this Guide, especially the later chapters, is on building and system details that can help achieve the desired results.

ENHANCED WORKING ENVIRONMENTS

Healthier working environments include favorable lighting, acceptable sound levels, and thermal comfort and are affected by many energy-efficiency measures. An increasing number of surveys and studies show that natural light, proper ventilation, appropriate temperature and humidity ranges, or even localized controls lead to healthier environments, increasing office worker productivity (Miller et al. 2009).

Daylighting, which uses natural light to produce high-quality, glare-free lighting, can improve office worker performance by as much as 25% (HMG 2003). Because it requires little or no electrical lighting, daylighting is also a key strategy for achieving energy savings. Quality lighting systems include a combination of daylighting and energy-efficient electric lighting systems. These complement each other by reducing visual strain and providing better lighting quality.

Advanced energy-efficient heating and cooling systems can produce quieter, more comfortable, and more productive spaces. Healthy indoor environments can increase employee productivity, according to an increasing number of case studies (Miller et al. 2009). Because workers are by far the largest expense for most companies (for offices, salaries are 72 times higher than energy costs, and they account for 92% of the life-cycle cost of a building), this has a tremendous effect on overall costs.

LOWER CONSTRUCTION COSTS/FASTER PAYBACK

Through use of an integrated design process and project delivery, energy-efficient offices can cost less to build than typical offices. For example, optimizing the building envelope for the climate can substantially reduce the size of the mechanical systems. An office with strategically designed glazing will have lower mechanical costs than the one without—and will cost less to build. In general, an energy-efficient office

- requires less heating and cooling energy use,
- costs less to operate,
- requires fewer lighting fixtures due to more efficient lighting systems,
- allows for downsized heating systems due to better insulation and windows, and
- allows for downsized cooling systems with a properly designed daylighting system, more efficient electric lighting, and a better envelope.

Some strategies may cost more up front, but the energy they save means they often pay for themselves within a few years.

REDUCED OPERATING COSTS

According to the most recent Commercial Buildings Energy Consumption Survey (CBECS) conducted by the U.S. Energy Information Administration (EIA), offices are ranked as the largest in terms of both floor space and primary energy consumption in the commercial building sector in the United States. Offices constitute 20% of the primary energy use and represent 19% of the total square footage of commercial building stock. By using energy efficiently and lowering an office's energy bills, hundreds of thousands of dollars can be redirected each year into upgrading existing facilities, increasing office workers' salaries, and investing in the latest technology in office appliances and equipment. Strategic up-front investments in energy efficiency measures provide significant long-term operating and maintenance cost savings. More importantly, studies show that productivity benefits are estimated to be as much as ten times the energy savings (Kats 2003).

REDUCED GREENHOUSE GAS EMMISSIONS

According to some estimates, buildings are responsible for nearly 40% of all carbon dioxide emissions annually in the United States. Carbon dioxide, which is produced when fossil fuel is burned, is the primary contributor to greenhouse gas emissions. Office buildings can be a part of the solution when they reduce their consumption of fossil fuels for heating, cooling, and electricity. The office occupants, staff, and society will appreciate such forward-thinking leadership.

REFERENCES

ASHRAE. 2004. ANSI/ASHRAE/IESNA Standard 90.1-2004, *Energy Standard for Buildings Except Low-Rise Residential Buildings*. Atlanta: American Society of Heating, Refrigerating and Air-Conditioning, Inc.

CoStar. 2008. Does green pay off? *Journal of Real Estate Portfolio Management* 14(4). www.costar.com/JOSRE/doesGreenPayOff.aspx, accessed January 20, 2011.

HMG. 2003. Windows and offices: A study of office worker performance and the indoor environment. Fair Oaks, CA: Heschong Mahone Group. www.h-m-g.com/downloads/Daylighting/order_daylighting.htm.

Kats, G., L. Alevantis, A. Berman, E. Mills, and J. Perlman. 2003. The costs and financial benefits of green buildings—A report to California's Sustainable Building Task Force. www.usgbc.org/Docs/News/News477.pdf. Sacramento: California Department of Resources Recycling and Recovery (CalRecycle).

Miller, N.G., D. Pogue, Q.D. Gough, and S.M. Davis. 2009. Green buildings and productivity. *Journal of Sustainable Real Estate* 1(1):65–89. www.costar.com/josre/JournalPdfs/04 Green-Buildings-Productivity.pdf.

RMI. 1994. *Greening the Building and the Bottom Line*. Boulder, CO: Rocky Mountain Institute. www.rmi.org/rmi/Library/D94-27_GreeningBuildingBottomLine, accessed January 20, 2011.

Introduction

1

Advanced Energy Design Guide for Small to Medium Office Buildings (AEDG-SMO; the Guide) provides user-friendly, how-to design guidance and efficiency recommendations for small to medium office buildings up to 100,000 ft^2. The intended audience of this Guide includes, but is not limited to, building owners, architects, design engineers, energy modelers, general contractors, facility managers, and building operations staff. Specially, Chapter 2 is written for a target audience of all design team members, whether they are design professionals, construction experts, owner representatives, or other stakeholders. Chapters 3 through 5 orient more toward design professionals to pursue sound design advice and identify interdisciplinary opportunities for significant energy reduction. Application of the recommendations in the Guide can be expected to result in small to medium size offices with at least 50% site energy reductions when compared to those same facilities designed to meet the minimum code requirements of ANSI/ASHRAE/IESNA Standard 90.1-2004, *Energy Standard for Buildings Except Low-Rise Residential Buildings* (ASHRAE 2004a).

This Guide contains recommendations to design a low-energy-use building and is *not* a minimum code or standard. A voluntary guidance document, this Guide is intended to supplement existing codes and standards and is not intended to replace, supersede, or circumvent them. The Guide provides both multidisciplinary design strategies and prescriptive design packages to significantly reduce energy consumptions in small to medium office buildings. Even though several design packages are provided in the document, this Guide represents *a way*, but *not the only way*, to build energy-efficient small to medium office buildings with 50% energy savings.

The energy savings projections of this Guide are based on site energy consumption rather than source energy. *Site energy* refers to the number of units of energy consumed on the site and typically metered at the property line. *Source energy* takes into account the efficiency with which raw materials are converted into energy and transmitted to the site and refers to the total amount of energy originally embodied in the raw materials. For example, it is generally accepted that site electrical energy is 100% efficient, but in fact it takes approximately 3 kWh of total energy to produce and deliver 1 kWh to the customer because the production and distribution of electrical energy is roughly 33% efficient.

This Guide was developed by a project committee that represents a diverse group of professionals and practitioners. Guidance and support was provided through a collaboration of members from American Society of Heating, Refrigerating and Air-Conditioning Engineers (ASHRAE), American Institute of Architects (AIA), Illuminating Engineering Society of North America (IES), U.S. Green Building Council (USGBC), and U.S. Department of Energy (DOE).

ASHRAE and its partners have, to date, published six Advanced Energy Design Guides (AEDGs) focused on new construction in small commercial buildings (ASHRAE 2010b). The purpose of these six published Guides is to provide recommendations for achieving at least 30% energy savings over the minimum code requirements of ASHRAE/IESNA Standard 90.1-1999 (ASHRAE 1999). Building types covered include small office, small retail, K-12 schools, small warehouses and self-storage buildings, highway lodging, and small hospitals and health-care facilities. The published guides are available for free download at www.ashrae.org/aedg.

GOAL OF THIS GUIDE

The AEDG-SMO strives to provide guidance and recommendations to reduce the total energy use in office buildings by 50% or more, on a site energy basis, using a building that complies with ASHRAE/IESNA Standard 90.1-2004 as the minimum code-compliant baseline (ASHRAE 2004a). The energy savings goal is to be achieved in each climate location rather than at an aggregated national average. The 50% savings is determined based on whole-building site energy savings, which includes process and plug loads.

One significant difference in this Guide is the inclusion of the Integrated Project Delivery (IPD) and multidisciplinary design strategies in addition to the prescriptive recommendations as in the previous 30% AEDGs (ASHRAE 2010b). It is very challenging to reduce energy use by half using a set of prescriptive recommendations because design solutions in a particular building vary depending on climate, site, building use, local code jurisdiction requirements, and other factors. Even though this Guide is intended to provide a variety of prescriptive design packages to offer architects and designers options, it also acknowledges that not all options will be appropriate solutions for an individual project. The climate-specific design strategies will provide guidance and flexibility to design professionals.

SCOPE

This Guide applies to small to medium office buildings up to 100,000 ft^2 in gross floor area. Office buildings include a wide range of office-related activities and office types, such as administrative or professional offices, government offices, bank or other financial offices, medical offices without medical diagnostic equipment, and other types of offices. These facilities typically include all or some of the following types of space usage: open plan and private office, conference meeting, corridor and transition, lounge and recreation, lobby, active storage, restroom, mechanical and electrical, stairway, and other spaces. This Guide does not consider specialty spaces such as data centers, which are more typically presented in large offices.

The primary focus of this Guide is new construction, but recommendations may be equally applicable to offices undergoing complete renovation and in part to many other office renovation, addition, remodeling, and modernization projects (including changes to one or more systems in existing buildings).

Included in the Guide are recommendations for the design of the building opaque envelope; fenestration; lighting systems (including electrical interior and exterior lights and daylighting); heating, ventilating, and air-conditioning (HVAC) systems; building automation and controls; outdoor air requirements; service water heating; and plug and process loads. Additional savings recommendations are also included but are not necessary for 50% savings. These additional savings recommendations are discussed in the "Additional Bonus Savings" section of Chapter 5 and provide design tips for additional daylighting (toplighting), natural ventilation, and renewable energy systems.

The recommendation tables in Chapter 4 do not include all the components listed in ASHRAE/IESNA Standard 90.1-2004. Though this Guide focuses only on the primary energy systems within a building, the underlying energy analysis assumes that all the other components and system comply with the minimum design criteria in Standard 90.1 and ANSI/ASHRAE Standard 62.1, *Ventilation for Acceptable Indoor Air Quality* (ASHRAE 2004a,

2004b). Also, Chapter 3 of this Guide provides general guidance and resources in terms of conditions to promote health and comfort.

In addition, AEDG-SMO is not intended to substitute for rating systems or references that address the full range of sustainability issues in office design, such as acoustics, productivity, indoor air quality, water efficiency, landscaping, and transportation, except as they relate to energy use. Nor is it a design text. The Guide assumes good design skills and expertise in office building design.

HOW TO USE THIS GUIDE

- Review Chapter 2 to understand how an IPD process is used to maximize energy efficiency. Checklists show how to establish and maintain the energy savings target throughout the project.
- Review Chapter 3 to understand the integrated design strategies, including architectural design features and energy conservation measures by climate zone. This chapter provides integrated design strategies for design professionals to make good decisions at the project early design stage. This is especially important when designers have to design a unique project on a specific site whose characteristics do not match the analyzed baseline building in shape, orientation, or glazing in the Guide.
- Use Chapter 4 to review climate-specific design strategies and select specific energy saving measures by climate zone. This chapter provides prescriptive packages that do not require modeling for energy savings. These measures also can be used to earn credits for the Leadership in Energy and Environmental Design (LEED) Green Building Rating System (USGBC 2011) and other building rating systems.
- Use Chapter 5 to apply the energy-saving measures in Chapter 4. This chapter has suggestions about best design practices, how to avoid problems, and how to achieve additional savings with energy-efficient appliances, plug-in equipment, and other energy-saving measures.
- Refer to the appendices for additional information:
  Appendix A—Envelope Thermal Performance Factors
  Appendix B—International Climatic Zone Definitions
  Appendix C—Commissioning Information and Examples
  Appendix D—Early-Phase Energy Balancing Calculations
- Note that this Guide is presented in Inch-Pound (I-P) units only; it is up to the individual user to convert values to metric as required.

ENERGY MODELING ANALYSIS

To provide a baseline and quantify the energy savings for this Guide, two office prototypical buildings were developed and analyzed using hourly building simulations. These building models include a 20,000 ft^2 small office and a 50,000 ft^2 medium office, each of which was carefully assembled to be representative of construction for offices of its class. Information was drawn from a number of sources and various office templates from around the country.

Two sets of hour-by-hour simulations were run for each prototype. The first set meets the minimum requirements of ASHRAE/IESNA Standard 90.1-2004 (ASHRAE 2004a), and the second uses the recommendations in this Guide. Each set of prototypes was simulated in eight climate zones adopted by International Code Council (ICC) and ASHRAE in development of the prevailing energy codes and standards. The climate zones were further divided into moist and dry regions, represented by 16 climate locations. All materials and equipment used in the simulations are commercially available from two or more manufacturers.

Energy savings for the recommendations vary depending on climate zones, daylighting options, HVAC system type, and office type but in all cases are at least 50% when compared to Standard 90.1-2004, ranging from 50% to 61%. Analysis also determined energy savings of approximately 46% when compared to Standard 90.1-2007 and 31% when compared to Standard 90.1-2010 (ASHRAE 2004a, 2007, 2010a). It is estimated that the energy savings are 55% using this Guide when compared to Standard 90.1-1999 (ASHRAE 1999), the baseline standard of the 30% AEDG series (ASHRAE 2010b). Energy-saving analysis approaches, methodologies, and complete results of the prototype office simulations are documented in two technical reports published by Pacific Northwest National Laboratory (PNNL): *Technical Support Document: 50% Energy Savings for Small Office Buildings* and *Technical Support Document: 50% Energy Savings for Medium Office Buildings* (Thornton et al. 2009, 2010).

ACHIEVING 50% ENERGY SAVINGS

Meeting the 50% energy savings goal is challenging and requires more than doing business as usual. Here are the essentials.

1. *Obtain building owner buy-in.* There must be strong buy-in from the owner and the operator's leadership and staff. The more they know about and participate in the planning and design process, the better they will be able to help achieve the 50% goal after the office becomes operational. The building owner must decide on the goals and provide the leadership to make the goals reality.

2. *Assemble an experienced, innovative design team.* Interest and experience in designing energy-efficient buildings, innovative thinking, and the ability to work together as a team are all critical to meeting the 50% goal. The team achieves this goal by creating a building that maximizes daylighting; minimizes process, heating, and cooling loads; and has highly efficient lighting and HVAC systems. Energy goals should be communicated in the request for proposal and design team selection based in part on the team's ability to meet the goals. The design team implements the goals for the owner.

3. *Adopt an integrated design process.* Cost-effective, energy-efficient design requires trade-offs among potential energy-saving features. This requires an integrated approach to office design. A highly efficient lighting system, for instance, may cost more than a conventional one, but because it produces less heat, the building's cooling system can often be downsized. The greater the energy savings, the more complicated the trade-offs become and the more design team members must work together to determine the optimal mix of energy-saving features. Because many options are available, the design team will have wide latitude in making energy-saving trade-offs.

4. *Consider a daylighting consultant.* Daylighting is an important energy savings strategy to achieve the 50% energy saving goal; however, it requires good technical daylighting design. If the design team does not have experience with a well-balanced daylighting design, it may need to add a daylighting consultant.

5. *Consider energy modeling.* This Guide provides a few design packages to help achieve energy savings of 50% without energy modeling, but whole-building energy modeling programs can provide more flexibility to evaluate the energy-efficient measures on an individual project. These simulation programs have learning curves of varying difficulty, but energy modeling for office design is highly encouraged and is considered necessary for achieving energy savings of 50%. See DOE's Building Energy Software Tools Directory at www.eere.energy.gov/buildings/tools_directory for links to energy modeling programs (DOE 2011). Part of the key to energy savings is using the simulations to make envelope decisions first and then evaluating heating, cooling, and lighting systems. Developing HVAC load calculations is not energy modeling and is not a substitute for energy modeling.

6. *Use building commissioning.* Studies verify that building systems, no matter how carefully designed, are often improperly installed or set up and do not operate as efficiently as expected. The 50% goal can best be achieved through building commissioning (Cx), a systematic process of ensuring that all building systems—including envelope, lighting, and HVAC—perform as intended. The Cx process works because it integrates the traditionally

separate functions of building design; system selection; equipment start-up; system control calibration; testing, adjusting and balancing; documentation; and staff training. The more comprehensive the Cx process, the greater the likelihood of energy savings. A commissioning authority (CxA) should be appointed at the beginning of the project and work with the design team throughout the project. Solving problems in the design phase is more effective and less expensive than making changes or fixes during construction. See the section "Using IPD to Maximize Energy Efficiency" in Chapter 2 and the section titled "Quality Assurance" in Chapter 5 of this Guide for more information. Appendix C also provides additional Cx information and examples.

7. *Train building users and operations staff.* Staff training can be part of the building Cx process, but a plan must be in place to train staff for the life of the building to meet energy savings goals. The building's designers and contractors normally are not responsible for the office after it becomes operational, so the building owner must establish a continuous training program that helps occupants and operation and maintenance staff maintain and operate the building for maximum energy efficiency. This training should include information about the impact of plug loads on energy use and the importance of using energy-efficient equipment and appliances.

8. *Monitor the building.* A monitoring plan is necessary to ensure that energy goals are met over the life of the building. Even simple plans, such as recording and plotting monthly utility bills, can help ensure that the energy goals are met. Buildings that do not meet the design goals often have operational issues that should be corrected.

REFERENCES

ASHRAE. 1999. ANSI/ASHRAE/IESNA Standard 90.1-1999, *Energy Standard for Buildings Except Low-Rise Residential Buildings*. Atlanta: American Society of Heating, Refrigerating and Air-Conditioning.

ASHRAE. 2004a. ANSI/ASHRAE/IESNA Standard 90.1-2004, *Energy Standard for Buildings Except Low-Rise Residential Buildings*. Atlanta: American Society of Heating, Refrigerating and Air-Conditioning.

ASHRAE. 2004b. ANSI/ASHRAE Standard 62.1-2004, *Ventilation for Acceptable Indoor Air Quality*. Atlanta: American Society of Heating, Refrigerating and Air-Conditioning.

ASHRAE. 2007. ANSI/ASHRAE/IESNA Standard 90.1-2007, *Energy Standard for Buildings Except Low-Rise Residential Buildings*. Atlanta: American Society of Heating, Refrigerating and Air-Conditioning.

ASHRAE. 2010a. ANSI/ASHRAE/IES Standard 90.1-2010, *Energy Standard for Buildings Except Low-Rise Residential Buildings*. Atlanta: American Society of Heating, Refrigerating and Air-Conditioning.

ASHRAE. 2010b. Atlanta: American Society of Heating, Refrigerating and Air-Conditioning.

DOE. 2011. Building Energy Software Tools Directory. www.eere.energy.gov/buildings/tools_directory. Washington, DC: U.S. Department of Energy.

Thornton, B.A., W. Wang, M.D. Lane, M.I. Rosenberg, and B. Liu. 2009. *Technical Support Document: 50% Energy Savings Design Technology Packages for Medium Office Buildings*, PNNL-19004. Richland, WA: Pacific Northwest National Laboratory. www.pnl.gov/main/publications/external/technical_reports/PNNL-19004.pdf.

Thornton, B.A., W. Wang, Y. Huang, M.D. Lane, and B. Liu. 2010. *Technical Support Document: 50% Energy Savings for Small Office Buildings*, PNNL-19341. Richland, WA: Pacific Northwest National Laboratory. www.pnl.gov/main/publications/external/technical_reports/PNNL-19341.pdf.

USGBC. 2011. Leadership in Energy and Environmental Design (LEED) Green Building Rating System. Washington, DC: U.S. Green Building Council. www.usgbc.org/DisplayPage.aspx?CategoryID=19.

Integrated Design Process

2

Chapter 2 provides an overview of a how an Integrated Project Delivery (IPD) process works. Successful outcomes of an integrated design process directly relate to potential heating, ventilating, and air-conditioning (HVAC) and lighting savings. Chapter 2 is written for a target audience of all design team members, whether they be design professionals, construction experts, owner representatives, or other stakeholders. To some technical team members, this chapter may seem to focus or be biased toward "architectural thinking" instead of presenting specific technical strategies for improving building system performance. This is intentional. One of the most important concepts of IPD is that some of the simplest, up-front architectural decisions about a building's form, orientation, and even architectural style can have the greatest long-term impact on HVAC or lighting efficiency. IPD encourages that the earliest design decisions take these issues into account through active project team discussion and collaborative dialog between all parties on a project team.

The breadth of analysis completed in support of this 50% Advanced Energy Design Guide has shown that there is a great deal of interconnectivity necessary between the critical decisions made by building envelope designers, HVAC engineers, lighting designers, and owner team members. Because of this finding, and in response to requests from the stakeholder member organizations, this design guide has specifically included this chapter as guidance on aspects of the IPD process. It acknowledges that collaboration and communication are key to designing holistic solutions that are necessary to achieving the Guide's stated 50% target or greater.

Under traditional design processes, energy use or a building's overall energy performance are not typically discussed in any meaningful way by engineers, architects, and building owners at a conceptual design level discussion. For IPD to truly be successful at obtaining energy savings in the range of greater than 50%, this tendency needs to change. All parties need to engage in much more detailed dialog earlier in the design process—using a team approach—taking the time to understand how their portion of the work affects the greater design of the whole project. Chapter 2 attempts to present the principles of IPD in this context.

Because some important design contributors may be nontechnical, this chapter is written in language accessible to the lay person so that everyone can understand the critical nature of the relationships and the process of decision making to support energy-efficient designs. It should be noted that Chapter 3 provides much greater technical detail with regard to overall multidisciplinary design strategies that might be pursued, and Chapter 5 provides how-to tips to explain the nuances of the application of particular components.

Chapter 4 provides prescriptive recommendations for critical component-level performance criteria that should result in the target 50% energy savings. As *a way* to achieve the 50% savings, the compiled lists of recommendations were obtained from an iterative energy analysis process that fine-tuned component-level performance and its aggregating influence on energy savings accrued from other disciplines.

PRINCIPLES OF INTEGRATED PROJECT DELIVERY (IPD)

Integrated Project Delivery (IPD) is a new concept for design and construction that uses a team approach to project management throughout all phases of the design process.

In a traditional system, the owner, designers, consultants, and constructors work on a fragmented basis, with success often based strictly on the profit received (to each) rather than the overall success of the final product. In IPD, all parties involved in the development of a building work together to achieve the common goal of maximizing multiple efficiencies of the building by having a more collaborative design and construction process. This approach is intended to not only save money but also provide higher-performing buildings than have been seen with the traditional approach.

The efficiency and quality of the design and construction are obtained through the following items, all of which result in reciprocal respect and trust:

- Early involvement of all design and construction team members
- Initially agreed-upon and documented common, desired goals
- Avoidance of adversarial protectionism
- Open communication
- Mutual cooperation toward the shared set of goals

A key element and differentiator of the IPD process is the formation of the project team as early in the project as possible, with team members that are committed to the success of the final product. Mutually agreed-upon metrics of performance, success, and corresponding compensation then align all project members with the fundamental goal of overall project success. This inclusiveness and buy-in then facilitate open collaboration and trust between the participants, which is paramount to IPD's success.

Cooperation hinges on participants working together as an integral team—either as participants who will be involved in the entire project from beginning to end or as key supporting participants who are included for advice pertaining to their specific areas of expertise as needed throughout the project's development. As stated in the American Institute of Architects (AIA) publication *Integrated Project Delivery: A Guide* (AIA 2007):

> Through the participation of all of the project participants in the decision making process, whether as a member of the decision making body or in an advisory role, the project benefits because the process allows all project participants to bring their expertise to bear on the issue at hand.

Figures 2-1 and 2-2 illustrate the interactions in traditional design team (Figure 2-1) and integrated design team (Figure 2-2) structures.

IPD is a method that conscientiously moves toward a more open and transparent agreement, focusing on mutual benefits and rewards by encouraging parties to converge on the project outcomes rather than their individual goals. Project goals must be defined and recorded early in the project. These goals should

- be defined by collaborative agreement between all key participants;
- be concise, so that no disagreement can arise due to individual interpretation;

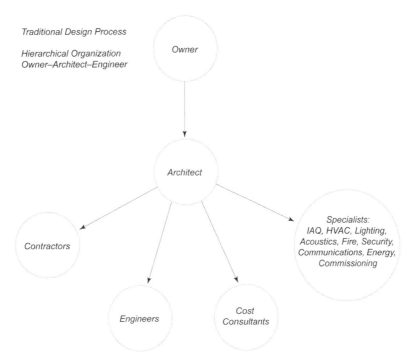

Figure 2-1   Traditional Project Design Team
Adapted from ASHRAE (2009)

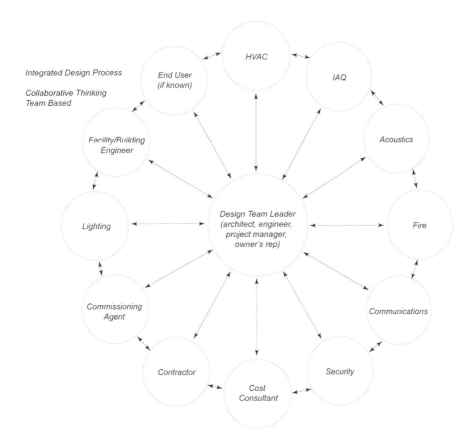

Figure 2-2   Integrated Project Design Team
Adapted from ASHRAE (2009)

- define congruent rewards based on individual risk (and include points that illustrate how all parties will benefit more due to the overall success of the project); and
- contain clauses that describe how problems are to be resolved—that they should be resolved "in the board room and not the courtroom."

At the heart of IPD is the promotion of communal discussions not only to bring multiple high-level expertise to bear on design but also to resolve problems or mistakes quickly and efficiently, avoiding finger pointing and accusations of liability. Decisions involving conflict resolution are handled internally by unanimous agreement from the project's decision-making body, avoiding the development of adversarial relationships between industry team members. Mutual benefit and trusting working relationships are IPD's driving forces toward better performance and more efficient buildings.

High-performance building design should be the focus for the future, and IPD will be the cornerstone for the evolution to more effective results in actual building performance. Additional details on how to set up and deliver an integrated project can be found in AIA's *Integrated Project Delivery: A Guide* (AIA 2007). A copy of the full guide can be downloaded from the contract documents section of the AIA Web site at www.aia.org/contractdocs/AIAS077630.

USING IPD TO MAXIMIZE ENERGY EFFICIENCY

As noted previously, IPD revolves around key collaboration agreements meant to remove barriers between parties and to encourage early contributions of wisdom and experience. While other sources are better suited to provide guidance on the contractual and organizational aspects, this section provides an idealized design management guidance template that describes how to use the more open relationships to achieve high energy-efficiency goals in the building design. Figure 2-3 gives a snapshot of the key steps in each phase that help lead toward energy-efficient solutions.

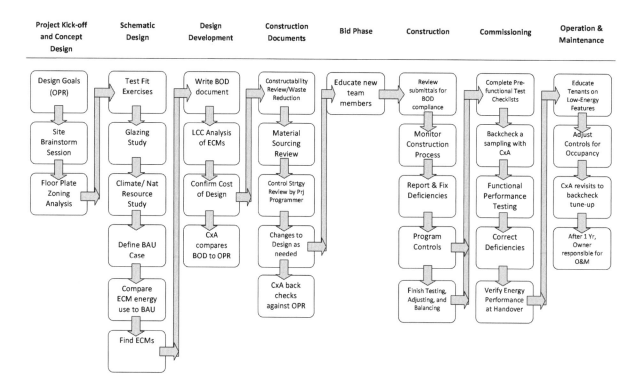

Figure 2-3 Key Design Activities for Energy Efficiency

DETAILS BY PROJECT PHASE

Project Kickoff

The project kickoff meeting is the most important meeting of the entire project with regards to establishing the Owner's Project Requirements (OPR). This exercise is usually led by the commissioning authority (CxA) and allows the owner's personnel and stakeholders to define what a successful project means for them. The requirements can cover construction costs, longevity, operating costs, specific characteristics, spaces required, functional aspects, specific maintenance or system preferences, frequency of use, aspirational goals, and any other critical issues. Functional requirements would typically include the indoor environmental quality aspirations such design criteria for available power density to support business processes, thermal comfort for occupants, acoustic performance of the space, lighting levels to support the necessary work tasks, and sufficient ventilation for indoor air quality (IAQ). It is strongly recommended that the traditional OPR arising from the commissioning process is further augmented to include the following information:

- Targeted energy labels and/or energy ratings—e.g., ENERGY STAR® (EPA 2011), Building eQ (ASHRAE 2011), Leadership in Energy and Environmental Design (LEED) (USGBC 2011), or Green Globes (GBI 2011)
- Payback period/return on investment thresholds used by the owner with regard to intelligent investment in construction upgrades
- Ownership/leasing arrangements, including utility back-charging or metering
- A clear prioritization of the requirements into "necessary for basic function," "necessary for intended function," and "desired upgrades"
- A clear prioritization of the topics within each of the categories above to guide future fund allocation decisions
- A clear delineation of the hierarchy related to decision making among the owner's constituent parties (who may request changes, who approves expenditures)
- Any pre-agreed-upon funding set-asides that are meant to achieve specific goals (such as those due to departmental contributions or a named donor)
- Any constraints imposed by the site, code, or planning agreements with the city, preexisting standards (if any), corporate sustainability policy statements, etc.
- Site-based measurements of actual plug-load usage of existing equipment or similar equipment at another owner-owned facility

The reason that this OPR information is necessary is that then all parties on the team are equally aware of the owner's priorities and can recommend systems that initially meet the stated criteria instead of spending time chasing alternates that cannot succeed. This lowers risk for all parties with regard to rework and provides a written document to refer to when budgetary pressures arise later in the design cycle. It is fully acknowledged that the OPR may grow over time to accommodate owner preferences that are only made apparent as the design progressed. Nevertheless, it is good practice to have a consolidated location for listing all original and added intent.

Concept Design

Concept design most successfully incorporates a series of free-flowing brainstorming sessions that allow the IPD team to review the OPR in the context of the site to look particularly for opportunities and risks. A key conceptual exercise usually covers a series of holistic site investigation and building configuration studies to look at which alternate best addresses the following issues:

- Status of site conditions (preexisting shading or wind-shadows caused from adjacent buildings or landscaping, outdoor air quality, outdoor ambient noise environment, site surface material)

- Availability of natural resources (sun, wind, geothermal energy, climate, bodies of water)
- Local material availability or reuse
- Site documentation with regard to storm-water runoff scheme and utilities available
- Status of surrounding buildings and review of code/planning regulations that may create obstructions to natural resources in the future or otherwise limit the design (such as the effects of the proposed building on its neighbors or other city planning concerns)
- Hardscaping or landscaping potential to reduce heat island effect or provide natural shading
- Security concerns
- Accessibility to transportation
- Sustainability opportunities
- Environmental risks or challenges

The goal of the design meetings is to quickly pass through a number of schemes and assess the pros and cons using only the expertise in the room. One cannot expect to link hard costs to each model in real time or to determine the final systems, but on the basis of past experience, the IPD team can rank each scheme qualitatively against the OPR. The exercise usually results in the entire team coming to some fundamental realization of the large-scale site constraints and identifying the three top architectural configuration alternates that might best meet the most fundamental of the OPR (usually square footage and first cost).

A secondary series of brainstorming sessions is often useful to look at the zoning of the floor plates that result from the building configuration exercise. This involves defining a 10–15 foot ring as a perimeter zone and the rest as an interior zone. A quick review of each façade orientation within the site constraints will reveal whether low-energy solutions are possible and what skin treatment or shading might be necessary to ensure comfort. The types of permutations to consider in this exercise include the following:

- Daylighting potential versus glare versus solar heat gain
- Reflectivity of other surfaces
- Natural ventilation potential for cooling
- Glazing types, shading devices, and fenestration size
- Operable window sizes (if allowed)
- Perimeter occupant comfort
- Projected heat loss/heat gain and impact on annual HVAC energy use
- Daylight harvesting and impact on potential energy savings
- Landscaping potential for natural shading

This exercise will quickly identify what proportion of the square footage has access to the outdoors through visual or physical connection. It is useful if the owner is able at that time to confirm the value of this connectivity as a priority for a positive and productive workplace environment.

Schematic Design

Schematic design is meant to give the IPD team time to further develop the ideas that still are feasible given the OPR and the constraints on site. At this phase, the architectural team members usually begin to identify where the various program occupancies are in the building as a "test-fit" exercise. Typical approaches generally place occupants at the exterior of the building for the outdoor connection mentioned above, leaving the center of the building for frequently unoccupied zones (conference rooms, copier rooms, storage areas, stairs, and restrooms) or high-heat-load areas (server rooms).

Very quickly this leads to the identification of façade locations where glazing would be desired to enhance the indoor environment for the occupants. At this point, the prediction of solar path (specifically, profile angle) is recommended to assess the exposure of intended glazing locations and the resulting penetration of solar rays into the occupied perimeter zone. It is

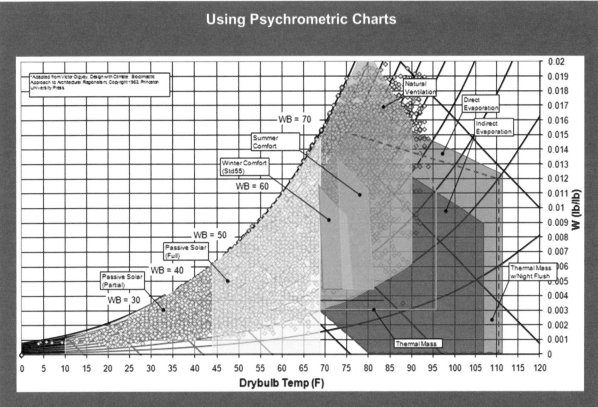

Using Psychrometric Charts

Frequency of Occurrence Psychrometric Chart
Source: Arup

In a frequency of occurrence psychrometric chart, each dot represents an hour in the year and its psychrometric state point. These charts provide a quick overview of the number of hours in which heating, cooling, humidification, or dehumidification are likely to be possible. Additionally, the graph superimposes a color scheme to identify those zones of outdoor air states in which more passive means of heating and cooling may be possible if sufficient numbers of hours fall into those ranges.

strongly recommended at this early phase that the design team also perform a back-check on glazing areas and thermal performance against the ASHRAE/IES Standard 90.1 prescriptive requirements (ASHRAE 2010) or the recommendations of this guide for the relevant climate zone. This will allow the IPD team to avoid getting too attached to a heavily glazed scheme that might fail the energy efficiency goals. Finding this out early will allow funds that might have been spent on increased glazing area to be more appropriately spent to improve the glazing performance or the overall building services performance to further reduce energy use.

Similarly, a more detailed weather/climate/natural resource analysis usually quantifies true "frequency of occurrence" potential for the following:

- Natural ventilation for cooling
- Free cooling through HVAC systems
- Daylighting
- Nighttime heat purge of thermal mass
- Heat recovery
- Use of radiant surfaces

All of this additional analysis begins to inform the design team with regard to what mechanical and electrical systems could be installed in order to provide a comfortable indoor environment. As a start, nominal mechanical and electrical plant room sizes, riser locations, and ceiling cross-sectional depths should be generated for the most traditional services approach, as this will serve as a base case for cost and life-cycle cost comparisons in subsequent phases.

Once a base-case building is created, it is costed and compared to the OPR to ensure that even the most standard of the available designs meets the first-cost and program requirements. If this is the case, it is often useful to perform a preliminary energy analysis by zone to determine approximate annual operating costs. This usually involves analyzing energy for one representative room per floor-plate zone (north, south, east, west, and interior) and projecting from that result an energy usage on a per-square-foot basis that can be extrapolated across all of the areas of similar characteristics. During schematic design, this level of calculation is usually good enough to confirm trends in energy savings associated with key design decisions. (See Appendix D for a more detailed description of this methodology to assist with this phase's preliminary energy analysis.)

The last necessary piece of work for the schematic design phase is to identify energy conservation measures (ECMs) that might be applied to the base case or to an alternate scheme that gives an identical performance compared to the base case. This is the point at which a thorough discussion of trade-offs is necessary to document, as choices have cascading effects on other systems. Typical exploratory interdisciplinary discussions during this phase include the following:

- Selection of structural material and its relative use as thermal mass or thermal insulation
- Selection of internal wall finish type and its potential to absorb internal heat gains
- Selection of floor material type and finish and its potential use as an air-distribution, heating, or cooling device
- Selection of façade type and orientation and each face's relative proportion and performance of glazing and opaque wall insulation
- Selection of glazing visible transmittance (VT) versus solar heat gain coefficient to allow daylighting without overheating
- Configuration of roofing shape/slope/direction and applicability of cool-roofing materials, clerestory skylights, and/or installation of photovoltaic or solar hot water panels
- Selection of electric lighting approach and zoning compatibility to accommodate ambient versus task lighting, occupancy sensors, daylight harvesting, and time-of-day controls
- Commitment to ENERGYSTAR equipment for plug-load usage reduction
- Review of plug-load use intensity by the owner's personnel
- Review of alternative HVAC and comfort cooling systems

As a conclusion to the discussions, the design team usually identifies a certain number of ECMs that they wish to pursue as optimizations imposed onto the base case. At that point, a fuller, more complex energy model is created to test the relative operating cost savings that are achieved for having invested first costs to install the upgrade. A matrix of options is usually developed to allow the IPD team to assess each ECM against a common set of criteria, usually including at least the following:

- Additional first-cost investment
- Anticipated annual energy cost savings
- Anticipated annual maintenance cost savings
- Simple payback period
- Return on investment (ROI)
- Reduction of $kBtu/ft^2/yr$
- Carbon emissions savings
- Additional percentage savings as compared to ASHRAE/IES Standard 90.1 (ASHRAE 2010)

- Potential additional USGBC LEED points for Energy and Atmosphere Credit 1 (USGBC 2011)
- Range of indoor temperatures achieved throughout the year
- Range of lighting levels achieved throughout the year

Obviously, there may be other project-specific OPR that should be incorporated into the matrix. The key point is that it is important for all parties to understand the whole view of any ECM application so that a balanced decision can be made, inclusive of all impacts on the desired goals of the project. The goal is to pick a selection of ECMs to pursue on a single scheme during the design development phase. This alignment of intent is quite crucial before significant design work and final calculations are commenced in the design development phase.

Design Development

The design development phase involves applying the final package of ECMs onto the unique architectural scheme for the project. The final energy models are usually used for submission to code authorities to show compliance with ASHRAE/IES Standard 90.1 (ASHRAE 2010) and may be used for submissions for LEED (USGBC 2011). The development of the scheme includes further design, calculation, and documentation of all building envelope, lighting, and mechanical/plumbing services that are regulated by code and owner-agreed limits on plug-load densities. Additionally, there is often a financial investment/life-cycle cost analysis (LCCA) of the ECM components in conjunction with the cost estimates, which are more detailed at this phase. The LCCA gives the entire team greater confidence in the value of the investment in all ECM upgrades at a point in the project where there are often pressures to engage in value engineering and other cost-cutting exercises. During this phase, the focus is primarily on documentation of design intent, and most disciplines with energy-using devices will write a Basis of Design (BoD) report that explains the full design and control intent. This BoD is then compared by the CxA to the OPR during a peer-review process to ensure that the owner's goals are likely to be met by the current state of design.

This phase usually provides a more detailed cost confirmation to ensure that the ECMs as applied have not adversely affected the construction budget. It should be noted that there are often value engineering discussions that occur during this phase, and items might be removed from the OPR or downgraded from "necessary" requirements in order to allow the project to move forward. If an ECM is removed or downgraded in this value-engineering phase, the unintended consequences of this decision must be explored; e.g., lowering the glazing VT to reduce solar heat gain may eliminate daylighting as an ECM, which will require investigation of additional ECMs to meet the stated energy-saving goals.

Because controls are essential to achieve energy efficiency at the levels proposed by this Guide, it is recommended that the design team produce a preliminary document describing the controls design intent for each category of HVAC and lighting equipment, including zoning approaches, required feedback, and energy-use monitoring, anticipated normal and emergency sequences of operations, and corrective algorithms to quickly return the system to an energy-efficient operation after it has been disrupted from stable operations. It is recommended that these sequences are reviewed by the CxA to ensure their feasibility in practical application.

Construction Documents (CDs)

The construction documents (CDs) phase is the final detailing of all systems, inclusive of sustainability features and ECMs. The mechanical, electrical and plumbing systems incorporate system drawings, specifications, BoD reports, controls drawings, controls points lists, and sequences of operation. The CxA usually reviews all of the documents and the updated BoD in order to confirm for the owner that the currently stated goals are on track for being met.

At this point in the design life, it is important for the construction team to confirm detailed constructability related to coordination, to review cost-optimization and waste-reduction techniques or sizing methods to be incorporated in the final drawings, and to confirm that they will provide the necessary documentation and acquisition of materials to meet the performance requirements for each of the energy-conservation measures. Additionally, the construction team should review the controls strategies and get clarification embedded into the specifications if there is an ambiguity about design intent.

The development of a simplified user-friendly tenant education guide by the design team is strongly recommended at this phase before the key designers and graphically capable team members demobilize at the change to the construction phase. The guide should be provided in a format that can be changed over the life of the building and should include just a few key items, such as the following:

- Sample introductory content for the landlord or highest-ranked on-site tenant manager to explain why energy efficiency is important to the business

- A "cartoon" type graphical representation of the energy-efficiency features of the building (a sample cartoon diagram showing energy-efficiency features is shown in Figure 2-4)

- A comparative estimate of maximum possible energy savings grounded in a comparison number that they are familiar with (a graphic example demonstrating possible savings to laypersons for a tenant education guide is shown in Figure 2-5)

- A "what you can do to help" section that lists the desired occupant behaviors

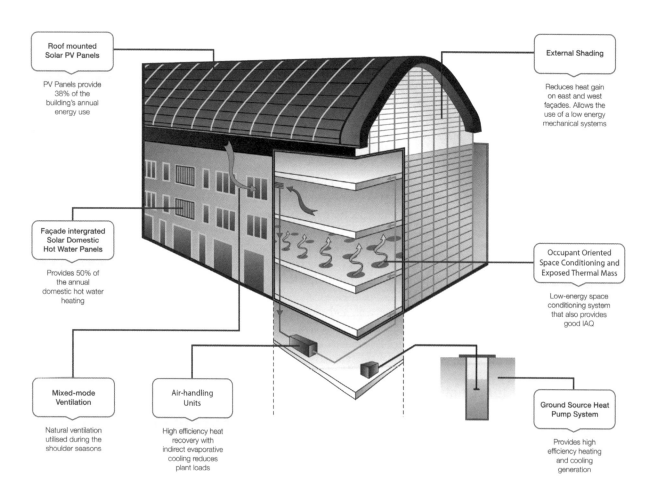

Figure 2-4   Example of "Cartoon" Diagram Describing Energy-Efficiency Measures
Adapted with permission from Arup

It is essential to have a fully documented controls package inclusive of points list, controls drawings, and sequences of operation for every controlled system in the building. This includes HVAC systems, lighting systems, and any plug-load setback or shutdown systems. The integration of a controls contractor at this point in the process is recommended to ensure that the intended sequences are capable of being programmed given the control system architectures available on the market. The designer, controls contractor, and CxA should review the intent in a collaborative method meant to ensure the success of the stable control of all components as an integrated system and identify any high-risk areas associated with implementation in the field.

Bid Phase

Usually on an IPD team, the construction team has been on board since the start of the project, so this phase is not relevant. If the IPD has been partial, and subcontractors in key phases are brought in at this phase, then it is necessary for the whole IPD team to educate the bidders on design intent and to stress the team's commitment to the ECMs that have been proposed. As the team grows exponentially at the construction phase, it is absolutely a necessity that all incoming parties are brought into the collaborative mind-set and are indoctrinated into the project principles, approaches, and commitment to a low-energy design.

Construction

During the construction phase, the CxA and the design professionals on the IPD team will review submittals to ensure compliance of the proposed materials with the required performance as stated within the CDs. The IPD team will ensure that items are installed to meet all regulatory requirements and are in compliance with the manufacturers' warranty and performance requirements. All parties are responsible to review the installation, report any deficiencies in installed work, and require remedial efforts to correct. Any deviation from the CDs must be documented with proof that the substitution will not adversely affect energy efficiency.

Near the end of the construction process, after all equipment is installed and the building is closed, equipment manufacturers will perform testing procedures during start-up and will confirm that the equipment is correctly operational. A testing, adjusting, and balancing contractor manipulates the settings on the equipment to achieve the correct water flows and airflows as required in the CDs.

The contracting team and the manufacturers' representatives are responsible for producing a set of operation and maintenance (O&M) manuals and to perform the specific hours of training for the owner's personnel. It is strongly recommended that the key technical facility operators of the building be brought on board at least a month before the prefunctional checklists are started so that they can familiarize themselves with the design intent and then can accompany

  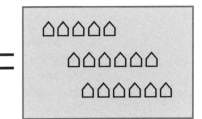

Baseline building in Miami
55.8 kBtu/sf/yr x 53,600 sf
= 2,990,880 kBtu/yr
= 876,540 kWh/yr

50% AEDG solution in Miami
25.6 kBtu/sf/yr x 53,600 sf
= 1,372,160 kBtu/yr
= 402,142 kWh/yr

Energy Savings equivalent
17 single-family homes at
27,841 kWh/yr
= 474,400 kWh/yr

Figure 2-5   Graphical Example Showing Comprehensible Scale of Energy Savings
Source: EIA (2009); U.S. total for 2005 was used

the contractor during the prefunctional checklist and functional testing protocols in order to familiarize themselves with the physical equipment.

Commissioning

The commissioning (Cx) process is the last performance testing applied to most new-build projects. The CxA will have written prefunctional checklists on the basis of the equipment submittals that were reviewed during the construction phase. The CxA will turn these checklists over to the construction team to fill in based on the manufacturers' start-up reports and other collected information such as warranty and wiring information. Once the prefunctional checklists are complete, the CxA will visit the site and do a random sampling to back-check a percentage of results to confirm that the reported findings are true and repeatable. Once this is confirmed, the CxA will release the functional test procedures, which were written in response to the contractor's detailed sequence of operations as submitted. The CxA will supervise the controls contractor running the equipment through its paces to prove adequate automatic reaction of the system in response to artificially applied inputs that simulate a variety of extreme, transition, emergency, and normal conditions. When this functional testing is complete to the satisfaction of the CxA, a report is written for the owner and handover can occur with confidence on the part of the owner that he or she has a fully functional building meeting the remaining requirements of the OPR.

Additionally, the CxA will usually assist with the supervision of the formal training of the owner's personnel in order to coach them on the appropriate corrective actions to take in the event that the system starts to drift away from its commissioned state. It is often useful if possible to run and monitor key aspects of the building for a one month period just before handover and to verify energy-related performance and the final setpoint configurations in the handover O&M documents. This will allow the owner to return the systems to the commissioned state (assuming good maintenance protocols) at a future point and have a set of comparative results.

Operation and Maintenance (O&M)

O&M of the equipment after handover are crucial to the energy efficiency for the life of the building. It is often the case that the first year after occupancy will reveal a truer nature of the real occupancy patterns within the building. This may be very different from the original design assumptions embedded in the energy model. It is recommended that the CxA have a contract with a service extension to return at the 12–18 month mark after handover in order to review the status of operations at that point in time and recommend slight adjustments to setpoints or modifications to the controls sequences in order to optimize the operation of the equipment with the view toward minimizing energy use. Occasionally, a second measurement and verification exercise is performed at this time in order to benchmark the energy use of each piece of equipment, usually in an alternate season from the original measurement and verification exercise if there are extreme seasonal differences in the particular climate.

Maintenance is the second owner-controlled aspect of operational efficiency. The O&M manuals will contain volumes of information about regular preventative maintenance, annualized maintenance activities, and periodic overhauls that should be performed to keep the equipment running at top performance. Just a few examples of how lack of attention to these matters can greatly reduce efficiency include: dirty filters increasing pressure drops, broken sensors causing poor feedback for the controls system, broken actuators disrupting demand-controlled ventilation (DCV) and/or air-side economizer cycles, and poor water quality fouling heat exchangers and inner surfaces of piping. Additionally, beyond the energy-efficiency issues, poor maintenance can lead to reduced performance. Examples include such items as poor IAQ arising from microbial growth occurring due to bad condensate drain pan maintenance, reduced indoor lighting levels due to depreciation of lamp output or dirt accumulation on lighting fixtures, and discomfort arising from lack of modulating control of supply air due to out-of-calibration sensors within the space.

Ongoing Tenant Education

In buildings designed for low energy use, it is always important to remember that occupant behavior is crucial to achieving the anticipated performance. When people understand the goals of the building and their roles, they can substantially reduce energy use. It can be beneficial for the design and construction team to host a lunchtime talk for the initial tenants and describe the building's design intent and sustainability features. This is the ideal venue to introduce the occupants to the tenant education guide described previously and to allow the local leadership to state their support for the energy-efficiency initiative in the building. It is important to forewarn occupants that although the building was commissioned, low-energy buildings sometimes take a season or two to reach their final optimized control strategies, so all occupants are invited to submit comments on performance as additional inputs for improving the operations. This type of personal engagement will not only encourage positive-impact behaviors with the initial tenants, but these first adopters can act as efficiency coaches for all future staff members.

REFERENCES

AIA. 2007. *Integrated Project Delivery: A Guide.* www.aia.org/contractdocs/AIAS077630. Washington, DC: American Institute of Architects.

ASHRAE. 2009. *Indoor Air Quality Guide: Best Practices for Design, Construction, and Commissioning.* Atlanta: American Society of Heating, Refrigerating and Air-Conditioning Engineers.

ASHRAE. 2010. ANSI/ASHRAE/IES Standard 90.1-2010, *Energy Standard for Buildings Except Low-Rise Residential Buildings.* Atlanta: American Society of Heating, Refrigerating and Air-Conditioning.

ASHRAE. 2011. Building eQ. www.buildingeq.com. Atlanta: American Society of Heating, Refrigerating and Air-Conditioning Engineers.

EIA. 2009. Residential Energy Consumption Survey, Table 5a. U.S. Residential Using Site Energy by Census Region and Type of Housing Unit, 1978–2005. www.eia.doe.gov/emeu/efficiency/recs_5a_table.pdf. Washington, DC: U.S. Energy Information Administration.

EPA. 2011. ENERGY STAR. www.energystar.gov. Washington, DC: U.S. Environmental Protection Agency.

GBI. 2011. Green Globes. www.greenglobes.com. Portland, OR: Green Building Initiative.

USGBC. 2011. Leadership in Energy and Environmental Design (LEED) Green Building Rating System. www.usgbc.org/Display Page.aspx?CategoryID=19. Washington, DC: U.S. Green Building Council.

Integrated Design Strategies

3

INTRODUCTION

Integrated design strategies are necessary to achieve an energy efficiency level of greater than 50% toward a net zero building from ASHRAE/IESNA Standard 90.1-2004 (ASHRAE 2004). As noted in Chapter 2, no single discipline alone can apply sufficient measures to achieve this level of energy efficiency. The burden must be shared, and a holistic approach must be understood. Use of the Integrated Project Delivery (IPD) model in Chapter 2 begins to remove barriers between disciplines, and this chapter is written to help teams not only stay focused on the key collective agreements but also pursue sound design advice on where the whole team might look to find real interdisciplinary opportunities for energy reduction. It should be noted that the focus of this chapter is on technical multidisciplinary strategies to achieve significant energy savings, but this is not meant in any way to downplay the importance of the functional requirements of the building owner, including operational performance, occupant comfort, and indoor environmental quality. It should also be noted that this chapter's recommendations remain at a strategic level (i.e., "This is something to consider")—for more details on how to implement any particular component, please refer to the how-to tips in Chapter 5.

This chapter includes the following guidance:

- Overview of design influences
- Building and site design features
- Energy conservation measures (ECMs)
- Multidisciplinary coordination for energy efficiency

As with the 30% Advanced Energy Design Guide (AEDG) series, there have been numerous iterative energy modeling analyses completed by Pacific Northwest National Laboratory (PNNL) to support the prescriptive recommendations in Chapter 4. While the recommendation tables in Chapter 4 provide recommended best practices for selecting components and should be reviewed by design teams, the runs are based on a rectangular building with unobstructed ribbon or punched opening glazing on all four sides. Because it is unlikely that this will be the shape or architectural design intent for every project, this chapter helps return control to an integrated team of design professionals to make good decisions for a unique project on a specific site, especially one whose characteristics do not match the baseline building in orientation

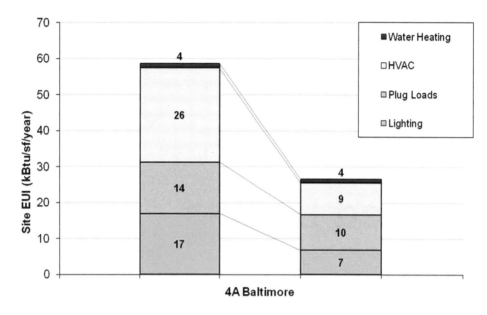

**Figure 3-1   Comparison of Baseline to Prescriptive 50% AEDG Solution
Showing Breakdown of Energy Savings Components**
Data source: Thornton et al. 2010

and glazing. Nevertheless, the analyses clearly show that systematically applied multidisciplinary approaches are essential to achieve the 50% energy savings.

Figure 3-1 shows the relative contribution of energy savings associated with each large-scale design component to build up to a 50% energy reduction over the ASHRAE/IESNA Standard 90.1-2004 baseline building. Within the heating, ventilating, and air-conditioning (HVAC) and lighting components reside key architectural decisions associated with the configuration of the façade. It should be noted that it was not possible in any of the energy modeling runs to achieve the 50% energy savings by looking at mechanical and lighting equipment optimization alone. Therefore, the performance of the envelope and its impact on loads and lighting become essential to reaching the energy use intensity (EUI) budget goal.

OVERVIEW OF DESIGN INFLUENCES

There are many design decisions that influence the energy use of a building (as expressed in EUI, or kBtu/ft^2·yr). The energy use of an office building is driven primarily by choices related to envelope, lighting, and HVAC systems. As a first step to reduce total energy use, the design team and owner should verify that the building infrastructure and equipment is sized based upon appropriate estimates of the proposed building usage, including optimized internal heat gains (e.g., high-efficiency lighting fixtures, ENERGY STAR®-labeled products) and the diversity and pattern of occupancy at the time of design.

The second area of design control is the building envelope, which involves the selection of building insulation and glazing to reduce heat transfer through surfaces, thereby reducing conduction and solar gains while enhancing daylighting opportunities. As noted in Figure 3-2, decisions about key elements of the building envelope are interrelated and heavily influence the heating and cooling strategies. It should be noted that while internal heat gains are all additive (i.e., cause the need for cooling), gains related to interaction with the outdoor climate, such as ventilation and building envelope, can be either heat gains or losses and are therefore heavily dependent on climate zone.

The third key area is that of reducing electrical energy consumption associated with lighting and plug loads.

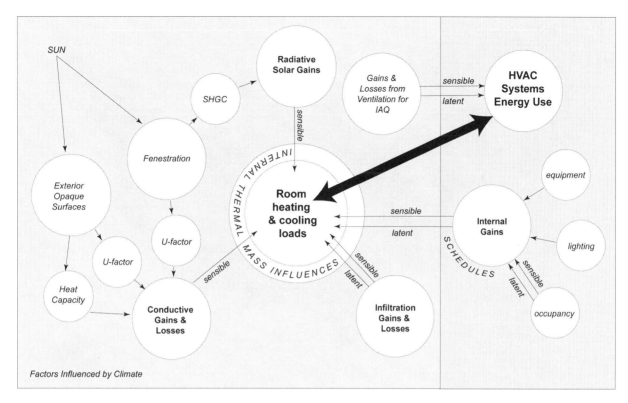

Figure 3-2   Heating and Cooling Influence

The rest of this chapter discusses each of the key concepts associated with minimizing the contribution of the heat inputs/losses and then summarizes the cooling and heating strategies most frequently applied by climate type as low-energy solutions. Guidance is provided that is related to the multidisciplinary design considerations prevalent when mixing and matching design decisions to achieve the optimal solution to meet the site constraints and architectural vision.

BUILDING AND SITE DESIGN FEATURES

There are many building architectural design features that impact the energy performance of a building. The major features include the building location (climate), shape, size, number of stories, and orientation. Each of these are presented in detail in this section.

CLIMATE FEATURES

Climate Characterizations by Location

There are several major climatic variables that impact the energy performance of buildings, including temperature, wind, solar energy, and moisture. These variables continuously change and can be characterized by annual or seasonal metrics.

- An indicator of the intensity and length of the heating season is represented by heating degree-days (HDDs), as shown in Figure 3-3 (DOI 1970).
- An indicator of the intensity and length of the cooling season is represented by cooling degree-days (CDDs), as shown in Figure 3-4 (DOI 1970).
- An indicator of the consistent intensity of the sun's energy is represented by the annual solar radiation, as shown in Figure 3-5 (DOI 1970).

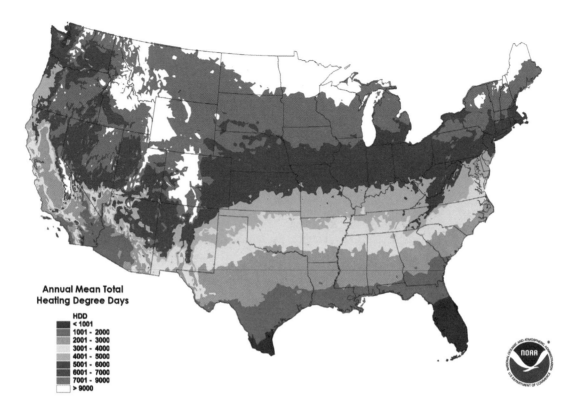

Figure 3-3   Heating Degree-Days
Source: NOAA (2005)

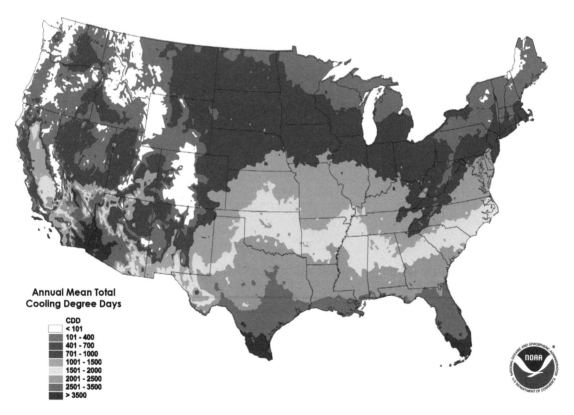

Figure 3-4   Cooling Degree-Days
Source: NOAA (2005)

Global Horizontal Solar Radiation - Annual

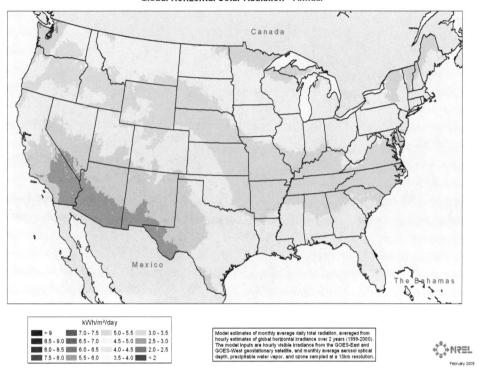

Figure 3-5   Annual Solar Radiation
Source: NREL (2005)

- An indicator of the worst case for removal of airborne moisture (i.e., dehumidification) is represented by the design dew-point temperature, as shown in Figure 3-6.

- An indicator of the ability of the air to engage in evaporative cooling is represented by the design wet-bulb temperature, as shown in Figure 3-7.

In combination, these variables show that distinct patterns emerge with regard to climate types, each of which has particular energy impacts on building design and operation. The U.S. has been divided into eight primary climate zones for the specification of design criteria in the major energy codes such as International Energy Conservation Code (ICC 2009), ASHRAE/IES Standard 90.1, and ASHRAE/USGBC/IES Standard 189.1 (ASHRAE 2010, 2009a). Figure 3-8 shows these climate zones as compared to CDDs and HDDs (Briggs et al. 2002a, 2002b, 2002c).

The characterization of these climate zones is based on seasonal performance metrics, not on the peak or design values. Each climate zone is clustered by HDD65 for the heating and CDD50 for the cooling; these climate zones are further subdivided by moisture levels into moist or humid (A), dry (B), and marine (C) to characterize their seasonal values. Sixteen cities have been identified as sufficient to represent all of the climate zones, as shown in Table 3-1 (CFR 1992). No single design strategy applies to all of these climate combinations. Each set of climate combinations needs to be analyzed separately.

It is important for the design team to determine the particular unique characteristics of the climate closest to the site. Annual hourly climate data is usually used for energy modeling and is available from federal government sources (EERE 2010). In addition to the acquisition of local data, it is necessary to assess any local topography or adjacent properties that would cause reduction in access to sunlight and passive solar heating.

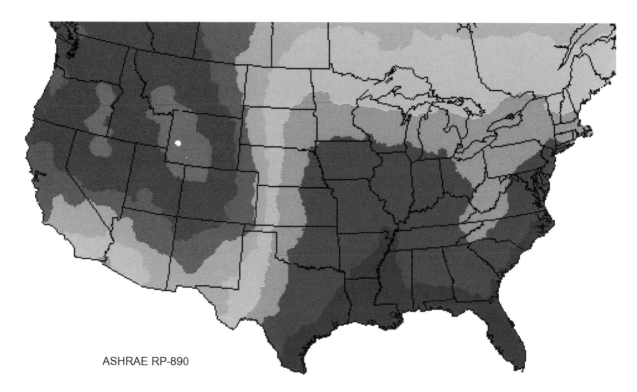

ASHRAE RP-890

Figure 3-6   Design Dew-Point Temperatures
Data source: Colliver et al. 1997

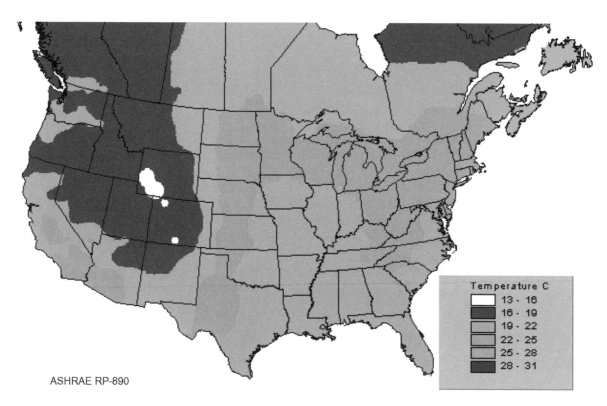

ASHRAE RP-890

Figure 3-7   Design Wet-Bulb Temperatures
Data source: Colliver et al. 1997

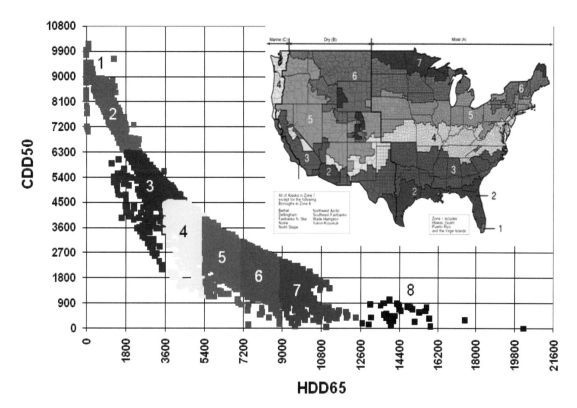

Figure 3-8 U.S. Climate Zone Map

Table 3-1 Cities Characterized by Climate Combinations

Climate	Hot	Mild	Cold	Very Cold	Extremely Cold
Marine		San Francisco - 3C Seattle - 4C			
Humid	Miami - 1A Houston - 2A Atlanta - 3A	Baltimore - 4A	Chicago - 5A Minneapolis - 6A		
Dry	Phoenix - 2B Los Angeles - 3B (coastal) Las Vegas - 3B (others)	Albuquerque - 4B	Denver - 5B Helena - 6B	Duluth - 7	Fairbanks - 8

Climate Influences

It is not reasonable to present every design strategy for each climate, but there are some fundamental principles that apply. The sensible and latent loads due to people are universal across all climate zones since the occupant densities and hours of occupation are assumed to be climate independent. Typically, the lighting power levels are the same but the energy use changes with location due to the daylighting available. Selection of the HVAC system is an important decision primarily because each system type has inherent efficiencies. Chapter 4 provides more information on specific climate zone strategies.

While there are benefits to the use of renewable energies (photovoltaics, solar, wind), these technologies are not design strategies that are required to achieve 50% energy savings. See the "Renewable Energy" section in the "Additional Bonus Savings" section of Chapter 5 for more information on these technologies.

Climate Dependence

Climate conditions are a major driving force, and there are multiple combinations that influence the energy performance of a building. Comparisons of the energies for heating, cooling, interior lights, exterior lights, plug loads, fans, pumps, and heat recovery in a medium office with radiant heating and cooling systems are shown in Figure 3-9.

A review of Figure 3-9 shows that there are distinct trends. In climates below 3000 HDD65, the cooling energy is greater than the heating energy. In climates above 5000 HDD65, heating energy use dominates over cooling energy use. In all climates the energy use is essentially constant for plug loads, interior lights, exterior lights, HVAC fans, pumps, and heat recovery. These relationships are similar for a medium office with a variable-air-volume (VAV) system; see Figure 3-10.

The heat released by the interior lights, plug loads, and fans add to the cooling load and diminish the heating load, which highlights the importance in addressing these loads in conjunction with the envelope constructions.

Fundamentally, what can be seen in Figures 3-9 and 3-10 is as follows:

- Lighting, plug, and fan loads are constant inputs and therefore are very consistent in the EUI budget. Indeed the only fluctuation most likely occurs from fan energy responding to on/off controls in response to climate.
- Heating EUI contribution increases with HDDs, as expected, but the scatter in the plot has to do with passive heating arising from solar contributions depending on the sunniness profile of the particular city. This becomes particularly obvious when looking at the pairs of heating and cooling contributions for a given HDD value—high heating goes with low cooling, which means that there is limited solar free heating. Similarly, high cooling goes with low heating, meaning there is a lot of solar heat to manage and the design team has to note whether the savings in heating sufficiently offset the penalties in cooling energy.

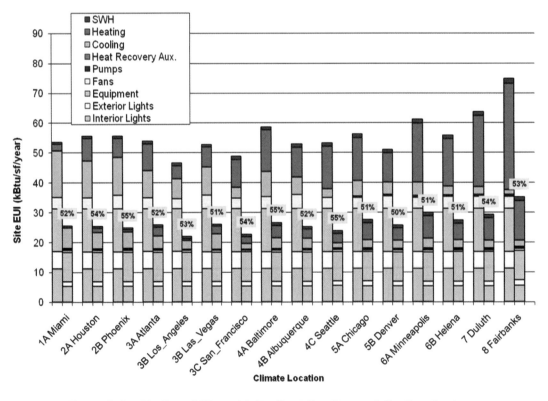

Figure 3-9 Medium Office with Radiant Heating and Cooling Systems

BUILDING FEATURES

Building Shape

The basic shape of the building has a fundamental impact on the daylighting potential, energy transfer characteristics, and overall energy usage of a building.

Building plans that are circular, square, or rectangular result in more compact building forms. These buildings tend to have deep floor plates that limit the potential of sidelighting a significant percentage of occupiable space. Building plans that resemble letters of the alphabet, such as H, L, and U, or that have protruding sections and surfaces at angles other than ninety degrees relative to adjacent building surfaces tend to have shallow floor plates where sidelighting strategies result in a higher percentage of daylighted floor area. (Atriums and other core lighting strategies may also be introduced into more compact building forms to achieve a similar effect.)

Less compact forms increase a building's daylighting potential, but they also may magnify the influence of outdoor climate fluctuations. Greater surface-to-volume ratios increase conductive and convective heat transfer through the building envelope. Therefore, it is critical to assess the daylighting characteristics of the building form in combination with the heat transfer characteristics of the building envelope in order to optimize overall building energy performance. (See DL2 in Chapter 5.)

The shape of the building also defines the window area and orientations that are available. Windows allow solar gains to enter the building; this is beneficial during the heating season but increases the cooling energy. The building shape needs to be designed so that the solar loading is properly managed. The solar management strategy changes by local climate characteristics, as solar intensity and cloudiness differ. Additionally, the shape of the building determines how wind impinges on the outdoor surfaces to assist natural ventilation or creates outdoor microclimates. In addition, attention must be paid to the effect of wind passing through openings in the

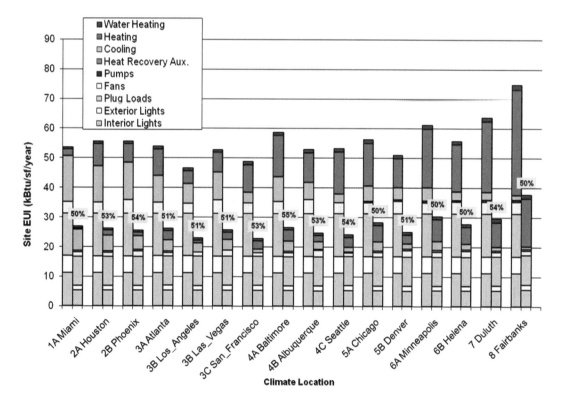

Figure 3-10   Medium Office with a VAV System

façade (e.g., windows, louvers, trickle vents, cracks), as this can drive unforeseen and/or uncontrollable infiltration.

Building Size

The size of the building impacts the energy use. Analysis of a 20,000 ft^2 two-story office building and a 53,600 ft^2 three-story office building clearly demonstrates the differences (Thornton et al. 2009, 2010). Figure 3-11 presents the baseline site energy use intensities for these two buildings for compliance with ASHRAE/IESNA Standard 90.1-2004 (ASHRAE 2004).

The size of the building also impacts the ECMs possible. For example, a small, 5000 ft^2 office building could be residential construction with wood-framed walls and ceilings as well as residential HVAC equipment in which the minimum efficiencies are set by the National Appliance Energy Conservation Act (CFR 1992). In these cases, there are limited options for obtaining more energy-efficient HVAC equipment.

Building size and especially depth of floor plate can have significant impacts on the feasibility of daylighting and natural ventilation.

Number of Stories

Typically, as the number of stories of a building increases, some aspects of design become more complicated. For instance, requirements for structural performance and durability/design life may affect choice of envelope components, the viability of exposed thermal mass, and the amount of area that may be used for fenestration. All of these may affect energy performance.

Taller buildings will have elevators with significant horsepower motors but intermittent energy use. Buildings requiring frequent vertical trips should consider the use of variable-

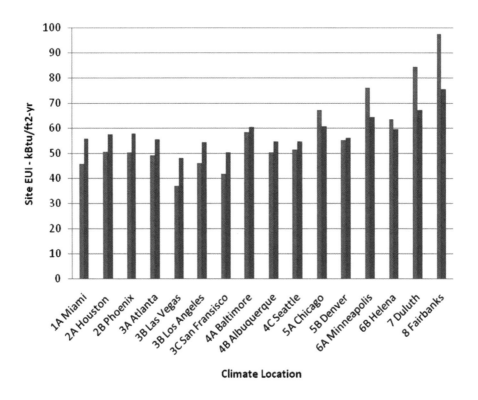

Figure 3-11   Site EUI for Office Buildings
Data source: Thornton et al. 2009, 2010

frequency drives (VFDs) on the motors and controls to stage the travel of elevators to reduce redundant trips in response to a call button.

Tall buildings run the risk of trapping a large block of space as purely internal and without connection to the outdoors. If an increased amount of space with access to natural daylight or ventilation is preferred, the designer of taller buildings can introduce toplight using skylights, clerestories, monitors, sawtooths, and atriums (See DL21 to DL27 in the "Additional Bonus Savings" section of Chapter 5). Horizontal glazing captures high-angle sun and may be difficult to shade. Exterior louvers, translucent glazing, vertical glazing, and other means should be considered to distribute toplight evenly into an interior space.

Building Orientation

The orientation of the office building has a direct impact on the energy performance primarily due to the orientation of the fenestration. The annual solar radiation impinging on a surface varies by the orientation and latitude, as shown in Figure 3-12 (Marion and Wilcox 2008).

The north solar flux (south solar flux for the southern hemisphere) is the least for any location; however, north daylighting is preferred due to no glare control requirements from direct sun penetration (reflections from adjacent buildings may require blinds on north windows for glare control). The east and west are essentially the same. The west exposure needs to be critically evaluated since it contributes to the peak or design cooling load. South-facing orientations in the northern hemisphere have the second largest solar intensity and the greatest variation in sun angle. Great care must be applied when designing external shading for this orientation, as attention must be paid to heat gain, glare, and the possibility of passive solar heating in cold climates. The horizontal solar flux is the largest and is critical if flat skylights on the roof are being considered. Clerestories on the roof facing north would be a preferred option.

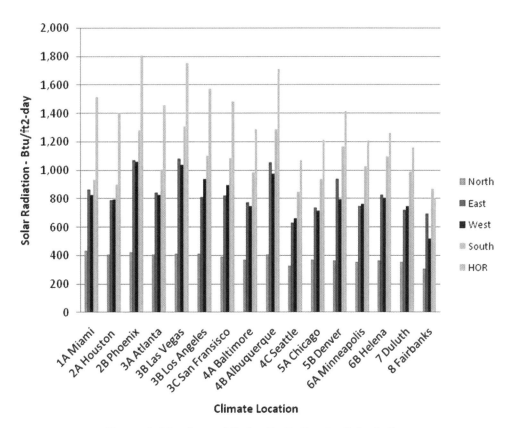

Figure 3-12   Annual Solar Radiation by Orientation
Data source: Marion and Wilcox 2008

Permanent projections can contribute to reducing the solar gains. Solar heat gain coefficient (SHGC) multipliers for permanent projections are presented in Figure 3-13. The largest energy reductions are on the south, east, and west orientations.

Building orientation and the placement of fenestration can have a significant effect on the ability of a design to provide useful daylight to perimeter zones. Using caution when doing simultaneous building configuration studies and internal space planning can maximize the amount of normally occupied space that can use daylighting for ambient light. For example, place all open office spaces on the north and south sides of the building where daylight is most easily managed.

Building Occupancy Types

Building function and occupancy type are key drivers for all internal heat loads. The density of the occupancy leads to heat from bodies, heat from equipment/computers, and heat from the electric lighting that runs in order to make the environment habitable. Some typical internal heat loads for office buildings are shown in Table 3-2 (see reference standards for area types not shown).

In addition to internal loads, other key occupancy-based criteria include the provision of sufficient ventilation to ensure indoor air quality (IAQ) in compliance with ASHRAE Standard 62.1 (ASHRAE 2010a). It is never the intent to achieve energy efficiency at the cost of human health. Additionally, depending on the function of work in the space, there may be acoustic design requirements similar to those stated in *ASHRAE Handbook—HVAC Applications*, Chapter 47, Table 42, "Design Guidelines for HVAC-Related Background Sound in Rooms" (ASHRAE 2007b, Table 42). Specific recommendations for lighting levels and visual contrast at and surrounding the work surface can be found in Chapter 4 of the tenth edition of the Illuminating Engineering Society of North America (IES) publication *The Lighting Handbook* (IES 2011).

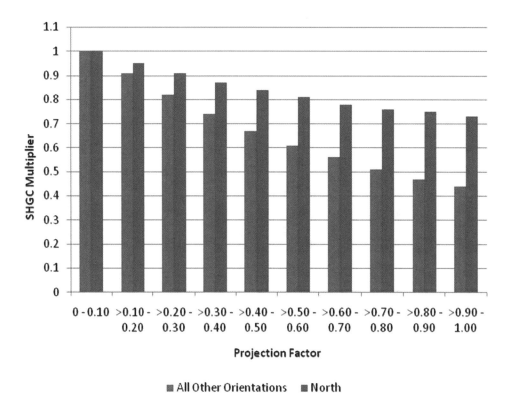

Figure 3-13 SHGC Multipliers for Permanent Projections
Source: ASHRAE (2007), Table 5.5.4.4.1

Building Orientation Considerations

Unshaded North Façade
Source: Arup

Armature for PV Panels
Source: Arup

Photographs of the California Department of Transportation building in Los Angeles show how the façade designer can tune each orientation for maximum daylight or solar protection. The north façade is unshaded with floor-to-floor glass. The south façade has an armature that holds photovoltaic (PV) panels at an angle to simultaneously shade the vision glass behind them and create electricity for the building. The eastern façade includes perforated metal panels that provide some level of shading for the glazed vision windows in the building envelope behind.

Table 3-2 Typical Internal Heat Gains for Office Spaces

Room Type	Occupancy Density, ft²/person*	Equivalent People Sensible Btu/h/ft²†	Lighting, Btu/h/ft² (W/ft²)**	Plug Load, Btu/h/ft² (W/ft²)‡
Office: light computer usage	200	1.22	3.75 (1.1)	1.7 (0.5)
Office: medium computer usage	200	1.22	3.75 (1.1)	3.4 (1)
Office: heavy computer usage	200	1.22	3.75 (1.1)	6.8 (2)
Conference room	20	12.25	4.4 (1.3)	1.7 (0.5)
Lobby	100	2.45	4.4 (1.3)	0.8 (0.25)
Corridor	—	—	1.7 (0.5)	0.8 (0.25)
Kitchenette/break room	20	12.25	4.1 (1.2)	1.7 (0.5)

*ASHRAE 2010a, Table 6-1.
†ASHRAE 2009b, Chapter 18, Table 1, applying "Seated, very light work" 245 Btu/h sensible.
**ASHRAE 2007a, Table 9.6.1.
‡ASHRAE 2009b, Chapter 18, Table 11.

ENERGY CONSERVATION MEASURES (ECMs)

The major ECMs focus on the envelope, lighting, plug loads, HVAC systems, and service water heating (SWH). This section of this chapter looks at each of these components to understand the relative design influence on total EUI.

The place to start is understanding where the baseline and business-as-usual buildings would start in terms of energy savings. An energy model can reveal the relative proportion of energy savings contributed by each design component. In general, most energy modeling programs output end-use data (i.e., by the component actually consuming the electricity or fuel) instead of linking the relative influence of design decisions directly to the output. Figure 3-14 shows a classic output that has all envelope and building configuration design decisions embedded in and diluted by total cooling, heating, interior lighting, exterior lighting, and HVAC system fan components. Figure 3-13 immediately identifies for the team the energy savings that should be the first point of attack in looking for ECMs. Clearly the heating and cooling energy savings vary significantly by location, so each requires particular attention specific to the climate. However, the energy savings from interior lighting, exterior lighting, interior equipment, and fans each contribute almost equally to the total energy savings, which means that these four major components all have to be addressed in every location.

The second key step to reducing energy use is to apply a series of ECMs, as noted in this Guide. Figure 3-15 shows an example of how one might use iterative energy modeling in a simplified approach to map the relative contributions toward energy savings achieved by each collective design decision made by the team.

The third key step is to identify packages of ECMs and understand the relative influence of key design decisions about those packages. Appendix D provides a method of using limited energy analysis in a small perimeter zone to help design teams understand the relative impact of their decisions on façade performance and building EUI. This appendix is provided so that teams can get "within striking range" with façade performance before moving to whole-building energy modeling by extrapolating total building energy use from knowing the worst-case conditions for each orientation. Design teams exceeding the limits of the early phase methods because of particularly complex geometries will have to perform full-scale whole-building energy modeling early in the design phase to inform design decisions.

ENVELOPE

The envelope is characterized by the opaque components and fenestration. Improvements should be considered for reduced thermal transmittance (i.e., U-factors), use of thermal mass, and control of solar heat gains.

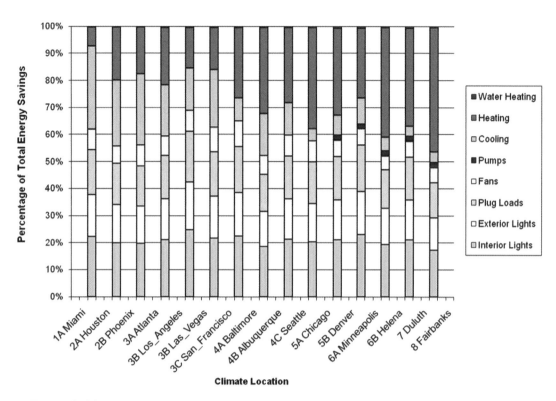

Figure 3-14   Percentage of Total Energy Savings arising from Each End-Use System
Source: Thornton et al. (2009)

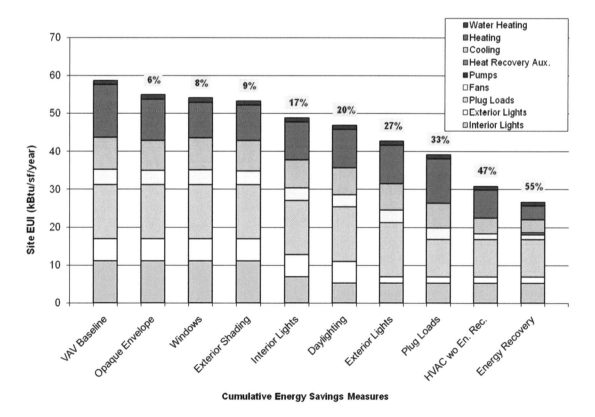

Figure 3-15   Relative Impact of Energy Savings Strategies

Upgrades to the opaque elements such as the roofs/ceilings (flat or sloped), walls (lightweight or mass), and foundations (slabs, crawlspaces, and basements) include increased insulation to lower thermal transmittance values (i.e., U-factors) or more thermal mass for roofs and walls. Adding cool roofs with high reflectivity in climates with more intense solar radiation is often found to be a direct benefit to reducing energy associated with cooling during the summer months.

The term *thermal mass* refers to the building's thermal capacitance, the amount of heat that is required to raise or lower the building temperature by a fixed amount. Greater thermal mass tends to reduce building peak conditioning loads by spreading these loads over an extended time span. Building thermal mass can reduce total conditioning loads if, across the daily cycle, exterior temperature conditions are both above and below the desired interior temperatures. Building thermal mass, furthermore, can absorb solar heat gain with reduced temperature rise and store that heat for later use for space heating after sunset. Similarly, during periods of lower humidity and low overnight temperatures, overnight ventilation with outdoor air (OA) can be used to cool a thermally massive building, offsetting subsequent daytime sensible cooling loads.

The fenestration has a major impact on both the architectural appearance and the energy savings potential. In addition, glazing provides daylighting and views for the occupants, connecting them to the outside world and improving occupant comfort and productivity (HMG 2003). Considerable effort needs to be focused on the fenestration designs to ensure the proper balance among heating, cooling, and daylighting is achieved. While electrical lighting energy can be saved through daylight harvesting, the other benefits of windows must be qualitatively weighed against the energy and cost-benefit analysis of increased HVAC energy usage due to larger window area or increased SHGC if implemented to improve access to daylight.

Orientation-sensitive window-to-wall ratios (WWRs) are recommended to help control solar heat gains while allowing more visible light at orientations where solar heat gains are not as much of an issue. Careful attention should also be paid to the issue of glare, a visual discomfort usually caused by the difference in relative brightness between a computer screen and a nearby window in direct low-to-medium-angle sun. Use of exterior shading such as overhangs on the south façade can also help control solar heat gains.

Other envelope design features to consider include the use of vestibules in order to reduce the introduction of OA through uncontrolled door usage. Additionally, placement and integrity of continuous air and vapor retarders is key to preventing the uncontrolled formation of condensation within the wall cavities, a situation that can lead to increased energy use to keep materials dry enough to reduce the risk of microbial growth and/or sick building syndrome. It is strongly recommended that a façade consultant, or person with similar expertise, be involved in detailing vapor retarder placement in low-energy buildings specifically because there will be less HVAC capacity available in base building systems to accommodate for a poor wall cavity construction. Careful façade detailing for sealing, especially at joints and fenestration interfaces, will also help reduce the amount of air leakage and infiltration experienced in the building—another potentially uncontrollable, continuous, real-time load on the HVAC system that can be mitigated with minimal amounts attention during the design and construction process.

In summary, the following approaches are often beneficial:

- Enhanced building opaque envelope insulation for exterior walls and roofs
- Use of mass in opaque envelope insulation to reduce cooling
- Inclusion of a cool roof in selected cooling-dominant climates with high solar intensity
- High-performance window glazing
- Exterior shading on south-facing windows
- Limited window areas at east and west
- Limited use of flat-roof skylights; consider north-facing clerestories/monitors
- Vestibules at openings to the outdoors
- Use of a continuous air barrier to reduce condensation risk and infiltration

LIGHTING

The lighting system is composed of three elements: daylighting, interior lighting, and exterior lighting. Daylighting can save electric lighting energy if there is a sufficient level of daylight available to meet interior illuminance requirements and if controls are employed to reduce the electric lighting in response to the available daylight. Interior lighting is a major energy user of the building and is constant across all climate zones; however, local site characteristics such as frequency of cloudy days or shading cast by adjacent properties will cause significant differences in overall interior energy use when daylighting is used. Equipment used and fixture energy density associated with exterior lighting are also consistent across all climate zones. It should be noted that energy use across climate zones with regard to annualized nighttime hours is constant; however, local site characteristics, such as which lighting zone (LZ) the building is located in, will cause differences in site energy use. Exterior LZs are a recognition of the types of surrounding buildings and are discussed later in this chapter in more detail.

Daylighting

Providing daylight is fundamental for an office environment, as it makes a key contribution to an energy-efficient and eco-friendly office environment. While the most valuable asset of daylight is its free availability, the most difficult aspect is its controllability, as daylight changes during the course of the day. Daylighting is more of an art than a science, and it offers a broad range of technologies that provide glare-free balanced light, sufficient lighting levels, and good visual comfort.

Daylighting strategies drive building shape and form, integrating them well into the design from structural, mechanical, electrical, and architectural standpoints.

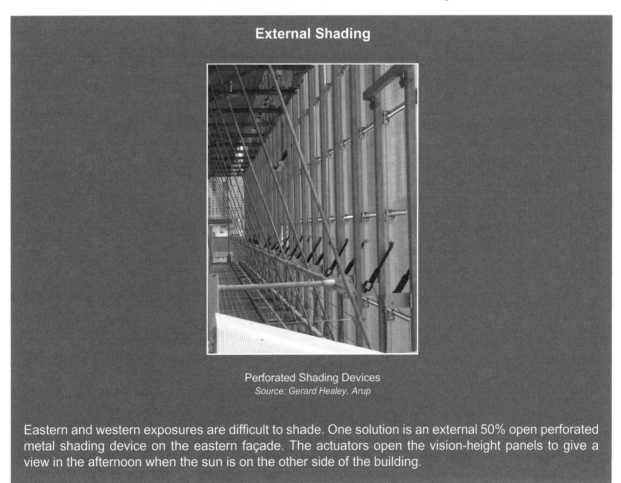

External Shading

Perforated Shading Devices
Source: Gerard Healey, Arup

Eastern and western exposures are difficult to shade. One solution is an external 50% open perforated metal shading device on the eastern façade. The actuators open the vision-height panels to give a view in the afternoon when the sun is on the other side of the building.

Daylighting increases energy performance and impacts building size and costs by downsizing fans, ductwork, and cooling equipment because overall cooling loads are reduced, allowing for trade-offs between the efforts made for daylighting and the sizing of the air-handling and cooling systems.

Daylighting will only translate into savings when electrical lighting is dimmed or turned off and is replaced with natural daylight. Effective daylighting uses natural light to offset electrical lighting loads. When designed correctly, daylighting lowers energy consumption and reduces operating and investment costs by

- reducing electricity use for lighting and peak electrical demand,
- reducing cooling energy and peak cooling loads,
- reducing fan energy and fan loads,
- reducing maintenance costs associated with lamp replacement, and
- reducing HVAC equipment and building size and cost.

However, to achieve this reduced cooling, the following criteria must be met:

- High-performance glazing is used to meet lighting design criteria and block solar radiation.
- Effective shading devices, sized to minimize solar radiation during peak cooling times, are used.
- Electric lights are automatically dimmed or turned off through the use of photosensors.

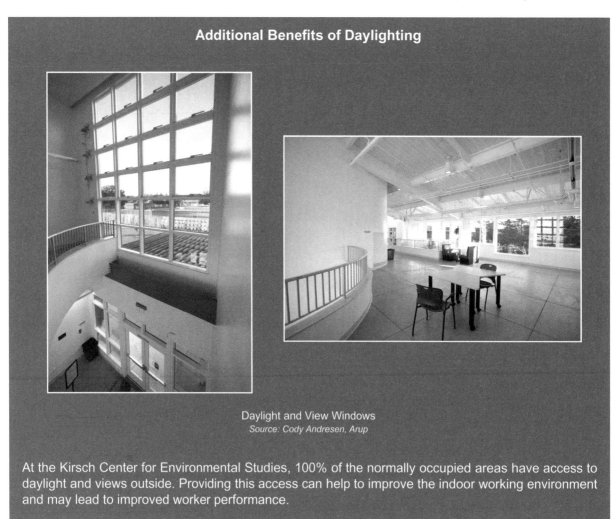

Additional Benefits of Daylighting

Daylight and View Windows
Source: Cody Andresen, Arup

At the Kirsch Center for Environmental Studies, 100% of the normally occupied areas have access to daylight and views outside. Providing this access can help to improve the indoor working environment and may lead to improved worker performance.

The case for daylighting reaches far beyond energy performance alone. Indoor environmental quality benefits the office workers' physical and mental health and has a significant impact on their performance and productivity. These impacts are difficult to quantify, but the potential for improvement and economical savings is immense and needs to be taken into consideration as serious decision-making criteria in the process of office design. These benefits may far outweigh the energy savings and become the significant drivers for daylighting buildings altogether.

The daylighting strategies recommended in this Guide have successfully been implemented in buildings. Most daylighting strategies are generic and apply to office buildings just as they do to other building types.

Interior Lighting

The primary lighting goals for office lighting are to optimize the open office spaces for daylight integration and to provide appropriate lighting levels in the private and open office spaces while not producing a dull environment. Producing a vibrant lighting environment is extremely important when attempting to minimize energy use, especially in the building's common areas (lobby, corridors, break rooms, and conference rooms). To achieve maximum lighting energy savings, lighting power densities (LPDs) need to be reduced and most spaces need to be provided with occupancy sensors and/or daylight-responsive dimming to reduce or shut off the lights when they are not needed. Additionally, the "night lighting," lighting left on 24 hours to provide emergency egress, needs to be designed to limit the power to 10% of the total LPD.

The interior space types typically found in office buildings are displayed in Table 3-3, along with the ASHRAE/IESNA Standard 90.1-2004 (ASHRAE 2004) percentage assumptions developed by PNNL. Each building space distribution will most likely be different, which creates different opportunities for energy savings. The building may have more than the standard 15% open office area, which creates more opportunities for daylighting, or the building may have more than the 29% private office area, which creates greater savings from occupancy sensors.

The first opportunity for energy savings is the reduction of the LPDs from those listed in the "Standard Baseline LPD" column in Table 3-3. Simple reductions are possible by using high-performance T8 lamps and premium low-ballast-factor (0.77) ballasts. This will reduce the T8 wattage by approximately 20% over the Standard 90.1-2004 LPD calculations.

Additional LPD savings are possible by using advanced luminaires such as the new two-lamp T8 or T5 high-performance lensed luminaires instead of prismatic lensed, parabolic, or recessed basket fixtures.

Table 3-3   Standard Percentage Assumptions by Space Type (Thornton et al. 2010)

Space Type	Percentage of Floor Area	Standard Baseline LPD, W/ft^2	Adjusted Baseline LPD, W/ft^2
Office—open plan	15%	1.1	0.68
Office—private	29%	1.1	0.8
Conference/meeting room	8%	1.3	0.77
Corridor/transition space	12%	0.5	0.5
Active storage area	14%	0.8	0.64
Restroom	4%	0.8	0.82
Lounge/recreation area	2%	1.2	0.73
Electrical/mechanical room	2%	1.5	1.24
Stairway	3%	0.6	0.6
Lobby	6%	1.3	1.09
Other	5%	1.0	0.82
Weighted LPD for the whole building	100%	1.0	0.75

Ensuring that lights are on only when someone is using them is an important opportunity as well. In Standard 90.1-2004 there are minimal requirements for occupancy controls. By adding manual ON or auto ON to 50% occupancy sensors to open office task lighting, general lighting in private offices, conference rooms, storage, and lounge/recreation spaces and auto ON occupancy sensors to electrical/mechanical rooms and restrooms, the lighting system will use 15% to 20% less lighting energy.

Exterior Lighting

Exterior lighting energy savings are accomplished by reducing the LPD and using automatic controls. Lighting power reduction recommendations are included in Chapter 5. Further savings are available by turning off façade lighting and reducing parking lighting by 50% between midnight and 6:00 a.m.

The new exterior LPD allowances in ASHRAE/IES Standard 90.1-2010 (ASHRAE 2010b) classify buildings in a five-zone lighting system (see Table 3-4). Very few buildings covered in this Guide are located in LZ4, which requires LPDs that are equivalent to those in Standard 90.1-2004. Most of the buildings addressed by this Guide will be located in LZ3, which requires approximately 35% less energy than LZ4. This 35% energy savings includes reducing the façade lighting allowance by 50%. Some smaller office buildings may be classified into exterior LZ2, which will reduce exterior lighting loads by approximately 50% over LZ4.

PLUG LOADS

Plug loads provide a significant opportunity for contributing to the overall building energy savings. Opportunities include strategies to reduce the connected power with more efficient equipment, substitute for lower power equipment, and reduce the amount of equipment and control equipment so that it runs at reduced power or is off when not in use.

Controlling plug-load energy usage is critical to achieve 50% and greater energy savings. Plug loads may use 15%–30% of typical office building energy, and that proportion tends to grow even higher with energy-efficient buildings as other uses such as lighting and HVAC are reduced. Plug equipment typically runs at normal operating power during occupied hours and in some cases may have the capability to partially power down when not in use. There is often potential to further reduce power during occupied hours when an office, cubicle, or other area with plug loads is not in use. Studies show that many types of plug-load equipment remain on at full power or reduced power even during unoccupied periods (Hart et al. 2004; Sanchez et al. 2007).

Two principal approaches are used: to select equipment with lower power demand and to control equipment so that it is on as little as possible when no one is using it. (See PL1 through PL6 in Chapter 5 for more detail. Note that control of task lighting is addressed in the lighting sections.) The approaches include the following tactics:

- Select lower-power equipment such as ENERGY STAR-rated equipment and other efficient equipment, and use laptop computers rather than desktop computers. Other concepts

Table 3-4 Exterior Lighting Zones

Lighting Zone	Description
0	Undeveloped areas within national parks, state parks, forest lands, rural areas, and other undeveloped areas as defined by the authority having jurisdiction
1	Developed areas of national parks, state parks, forest lands, and rural areas
2	Areas predominantly consisting of residential zoning, neighborhood business districts, light industrial buildings with limited nighttime use, and residential mixed-use areas
3	All other areas
4	High-activity commercial districts in major metropolitan areas as designated by the local jurisdiction

Source: ASHRAE (2010b)

include reducing the number of pieces of equipment in use, such as by consolidating printing services (Lobato et al. 2011).

- Control equipment so that it is off when not in use. Options include occupancy-sensor-controlled plug strips or outlets, occupancy-sensor-controlled vending machines, timer switches for equipment that is shared during occupied hours but can be off during unoccupied hours, and power management of computers and other devices, ensuring that sleep modes are fully active. Use of efficient low-voltage transformers and newer power management surge protectors can reduce phantom loads associated with low-voltage equipment (Lobato et al. 2011).

SERVICE WATER HEATING

As with other building services systems, the first step toward reducing the energy consumption of SWH systems is to reduce the loads on the system. For service hot water, technologies for load reduction include sensor activators and aerators for lavatory faucets, ultralow-flow showerheads, and ENERGY STAR dishwashers for break rooms. (See WH2 in Chapter 5.)

Office buildings typically have very low but often widely distributed service hot water loads. Restrooms, break rooms, and janitor closets typically are located on every floor. The performance requirement for the system, furthermore, is delivery of hot water to the fixture in a few seconds, requiring constant hot-water availability only a few feet from each fixture. For a one-floor office building with wet spaces clustered in the building core, the goal of rapid delivery of hot water to fixtures may easily be met. Multistory buildings and multicore buildings may present significant challenges.

Two strategies are typically used to achieve this goal of hot-water availability: positioning hot-water heaters immediately adjacent to each end-use location and utilization of a pumped recirculation loop to maintain elevated water temperatures in the distribution system across periods of nonuse. Each of these strategies has benefits and deficiencies.

Positioning the water heater adjacent to loads often requires multiple water heaters to service building loads, precluding utilization of renewable or "free" heat available from a single source. Examples of reduced-energy heating sources include solar domestic water heating and desuperheater heat recovery from direct expansion (DX) cooling units. In some code jurisdictions, gas supply piping and flue runs internal to the building may present compliance or cost challenges.

The recirculation loop strategy has the disadvantage of increased energy consumption due to both the pump energy and the increased heat losses through the distribution piping. Typical office buildings have a relatively small usage of hot water, and the heat loss through the distribution piping may entail more energy consumption than does the actual hot water use. Selection of the appropriate strategy should weigh all aspects of the two approaches.

HVAC SYSTEMS

A variety of HVAC systems can be used for small to medium office buildings up to 100,000 ft^2, all of which can meet the moderate heating and cooling loads and provide the minimum required outdoor airflow for ventilation. In general, the following actions should be considered when selecting low-energy HVAC systems:

- Reduce loads significantly as described in other sections of this Guide.
- Zone perimeter areas separately and consider local systems with increased cooling and heating capacities for spaces with significant glazed area. Perimeter convective heating systems should be considered as a means to offset heat loss from the area immediately adjacent to the façade. It is important to integrate and properly control perimeter and central systems to avoid simultaneous heating and cooling. Fan-coil units are often placed in perimeter zones to respond quickly to changes in load arising from heavily glazed surfaces.

- Decouple space ventilation and dehumidification from sensible conditioning, if beneficial, by independently conditioning and supplying OA. A 100% outdoor air system (100% OAS), often called a *dedicated outdoor air system* (DOAS), is often used to provide filtration, heating, cooling and dehumidification, and humidification of OA for ventilation. This Guide uses *100% OAS* and *DOAS* interchangeably. A 100% OAS may include energy recovery. These systems also reduce energy associated with dehumidification by eliminating or nearly eliminating energy for simultaneous cooling and reheating. They support the opportunity for cost-effectively applying energy recovery with a single energy recovery unit rather than a separate energy recovery unit for each system. For zonal systems such as air-source heat pumps and fan-coils, they allow the supply fans on the zonal units to cycle rather than run continuously (occupied hours) as they would when providing ventilation.
- Require designers to make a fundamental strategic decision as to whether they will pursue a centralized heating/cooling system or a more distributed approach where cooling and heating equipment reside physically close to the zone. Design options include the following:
  a. When lower operating costs and central maintenance and control are primary considerations, designers tend to select centralized systems that colocate DX units or chillers, boilers, and cooling towers (as applicable) away from occupied spaces. This also mitigates noise impacts. These systems often require a greater first cost than distributed systems but often result in annual energy savings because additional benefits include the ability to reduce installed capacity by using load diversity. Centralized cooling systems tend to be more cost-effective when the total building load exceeds 100 tons, depending on climate and patterns of occupancy use. Central boiler systems are applied in many building sizes since they can provide better close space control and can be used with many terminal unit types.
  b. When low first cost and simplicity are primary concerns, designers tend to select zone-by-zone distributed systems incorporating both heating and cooling capacity. This approach tends to be used for smaller buildings or larger buildings with sufficient roof area. Distributed equipment usually consists of fan, cooling coil, compressor, and outdoor condenser. Examples of distributed systems include packaged rooftop air conditioners and heat pumps as well as refrigerant-based split-system fan-coil units (single or multiunit). Water-source heat pumps (WSHPs) are also distributed systems in that the compressor is located close to the occupied space, but they are served by a centralized water system with auxiliary boilers and heat rejection devices.
- Make minimizing energy used to condition ventilation air a key goal. Strategies include demand-controlled ventilation (DCV) and energy recovery. DCV approaches for single-zone systems are well established; DCV for multiple-zone recirculating systems is more complicated and approaches are still evolving.
- Take advantage of the moisture- and heat-absorbing capacities of the OA. Strategies include providing economizers in all climate zones except climate zone 1, evaporative cooling and evaporative condensers in drier climates, and natural ventilation.
- Control fan and pumping energy for the selected system type. Pressure drop and friction should be reduced as much as possible for all fan and pump systems. Proper pipe and duct layout helps reduce friction. VFDs, even where not required by code, should be considered in order to match airflow and water flow (as necessary) to the loads within the systems. Controls using pressure reset on the basis of measured feedback should be considered on systems with variable flow.
- Select efficient equipment for all of the recommended HVAC systems. Target minimum efficiency values are included in Chapters 4 and 5.
- Note that radiant cooling systems (embedded in floors or ceilings) require separate control of humidity and the DOAS performs the main dehumidification function. Additionally, supply water temperature must be carefully monitored against space dew-point temperature to avoid condensation. Although this is not specifically included in the recommendations, design teams could evaluate the energy performance of passive chilled beams as another alternative consistent with the use of a DOAS and chilled-water supply.

Radiant Floor System

PEX Tubing in Serpentine Pattern
Source: Cole Roberts, Arup

For radiant floor systems, cross-linked polyethylene (PEX) tubing is placed in a serpentine pattern over the structural slab prior to the topping slab being poured. Density of tubing is related to heat output. Floor areas can be zoned, with thermostatic control at a central manifold; however, radiant heating and cooling systems tend to have a long start-up time related to the thermal mass of the surrounding concrete. Additionally, radiant heating and cooling floors tend to have a floor finish that is monolithically integrated with the topping slab in order to provide continuous conduction from the tubing. Carpet and other insulated finishes should be avoided.

- Note that VAV systems incorporating low-temperature air, energy recovery, DCV, and ventilation and supply air temperature reset (as shown in the recommendation tables in Chapter 4) significantly reduce HVAC system energy use.
- Consider water-based heat rejection systems, including WSHPs and ground-source heat pumps. Use of 100% OASs with energy recovery should be considered.
- Note that packaged heat pumps with DOASs offer a low first cost solution appropriate to smaller office buildings.

HVAC CONTROLS

HVAC control systems should achieve the following goals:

- Accurate provision of space conditions required for human comfort, health, and functionality in each space.
- Avoidance of conflicting operation of different parts or components of the conditioning systems (fighting).
- Optimizing operation of HVAC components most efficiently to meet required conditioning loads, especially during part-load conditions.
- Sequencing of HVAC components to meet required interior environmental conditions, including deactivation during unoccupied periods.

- Avoidance of excessive cycling of systems components and resultant accelerated wear and inefficiency.

For more information on specific HVAC systems, please see the how-to tips in Chapter 5.

QUALITY ASSURANCE

Quality and performance are the result of intention, sincere effort, intelligent direction, and skilled execution. A high-quality building that functions in accordance with its design intent, and thus meets the performance goals established for it, requires that quality assurance (QA) be an integral part of the design and construction process as well as the continued operation of the facility.

Deficiencies in the building envelope and mechanical and electrical systems have a wide range of consequences, including elevated energy use or underperformance of the energy-efficiency strategies. These deficiencies are commonly a result of design flaws, construction defects, malfunctioning equipment, or deferred maintenance. The QA process typically referred to as *commissioning* (Cx) can detect and remedy these types of deficiencies.

As facilities search for higher efficiency through innovation, new applications, and complex controls, the risk of underperformance and the potential for more deficiencies increases. To reduce project risk, Cx requires a dedicated person (one with no other project responsibilities) who can execute a systematic process that verifies that the systems and assemblies perform as required. The individual responsible to provide this is called the commissioning authority (CxA).

Success of the Cx process requires leadership and oversight. CxA qualifications should include an in-depth knowledge of mechanical and electrical systems design and operation as well as general construction experience. The individual represents the owner's interests in helping the team deliver a successful building project. The CxA can be completely independent from the project team companies or a capable member of the contractor, architect, or engineering firms. The level of independence is a decision that the owner needs to make.

The Cx process defined by ASHRAE Guideline 0, *The Commissioning Process* (ASHRAE 2005), and ASHRAE Guideline 1.1, *HVAC&R Technical Requirements for The Commissioning Process* (ASHRAE 2007c), is applicable to all buildings. Owners, occupants, and the delivery team benefit equally from the QA process. Large and complex buildings require a correspondingly greater level of effort than that required for small, simpler buildings.

Cx needs to be an integral part of the design and construction process in order to reach the energy performance goals a building requires. Chapter 5 contains descriptions of the recommended steps in the building design and construction Cx process. Appendix C provides examples for the Cx process, and more detailed information can be found in ASHRAE's Guideline 0 and Guideline 1.1.

MULTIDISCIPLINARY COORDINATION FOR ENERGY EFFICIENCY

OVERVIEW

Integrated design strategies require significant multilateral agreement on design intent from a variety of stakeholders. The following multidisciplinary recommendations are provided to identify a series of items for which a direction and agreement must be achieved. These are very different ideas as compared to the component-level or process-level discussions inherent in Chapter 5. Truly holistic low-energy design solutions are not achieved solely through the optimization of each component, but rather by exploiting the mutually beneficial synergies between otherwise independent design strategies.

MULTIDISCIPLINARY RECOMMENDATIONS

Define Business as Usual and Baseline Buildings

One of the very first things that the design team must define is what the "business as usual" (BAU) design solution would be. This will often be a minimally prescriptive ASHRAE/IES Standard 90.1 (ASHRAE 2010b) equivalent consisting of a square or rectangular building virtually filling the site with as low a profile as possible. The energy use of this building typically represents the high-end of allowable energy use and sets the comparative standard against which absolute savings are achieved on the road toward net zero energy use. As each ECM is applied, the design team should keep track of all incremental victories achieved at each step. The comparison to the BAU benchmark is a real measure of success reflective of all design decisions.

The second key item that the design team must define is what the baseline design solution would be once the preferred building configuration's design is completed. The baseline is very different from the BAU benchmark because the current ASHRAE/IES Standard 90.1 requires all proposed and baseline energy models to have identical shapes, footprints, and occupancies (ASHRAE 2010b). Thus, the baseline does not reward fundamental building configuration decisions for their positive effect on energy use.

It is important for the design team to agree to move away from both the BAU benchmark and the baseline in making proactive design decisions. It is also important that there is no shifting benchmark of success.

Benchmarking

While the BAU benchmark represents the highest allowable energy-use intensity on site by calculation methods, there are a series of other energy-use benchmarks that represent the existing building stock in the United States:

- U.S. Environmental Protection Agency and U.S. Department of Energy's ENERGY STAR Portfolio Manager (EPA 2011)
- U.S. Energy Information Administration's Commercial Buildings Energy Consumption Survey (CBECS) (EIA 2011)
- California Energy Commission's California Commercial End-Use Survey (CEUS) (CEC 2008)

It is possible to benchmark the proposed design against the BAU benchmark and against its preexisting peers to demonstrate that substantial steps have been taken toward energy-use reduction. Designers often successfully compare their designs to the typical equivalent building in the preexisting stock or to the number of houses that could be powered on the energy savings to make it easier for laypeople to understand the magnitude of the energy savings.

Historic data, however, is not the inspiration for good design in the future. This is where more aspirational benchmarking can benefit the team. The most frequently used benchmarks are the following:

- Energy savings as designated by percentage annual cost savings as compared to Appendix G of ASHRAE/IES Standard 90.1 (ASHRAE 2010b) (typically used by codes and policies, also used by the Leadership in Energy and Environmental Design [LEED] Green Building Rating System [USGBC 2011])
- Absolute EUI definitions (occasionally used by campuses, regularly used by the General Services Administration; easiest to measure and verify after construction)
- Net zero energy definitions

As noted above, it is important for the design team to agree to move away from the design practices that led to older poor-performing buildings and toward a quantifiable target that is consistent with the available funding for the job.

CMTA Office Building—A Case Study

CMTA Properties designed and built a new 20,000 ft^2 office building in Louisville, Kentucky, in 2008. The goal was to create a showcase of green, energy-efficient building design that could provide a living demonstration of these technologies for clients and would be the first step toward a net zero energy building.

Thermal Envelope

The exterior of the single building is designed to appear as three separate buildings in order to fit into the neighborhood's historical architectural style. The exterior walls are constructed of insulated concrete forms, a structural wall system that provides an excellent thermal barrier and reduces air infiltration. The full thermal envelope, which includes roof insulation and high-performance glass, exceeds ASHRAE/IESNA Standard 90.1-2004 by 20%. The building faces west and includes large front windows with 7 ft awnings for the first-floor conference rooms. The awnings provide solar and glare control.

Lighting/Daylighting

Strategies employed to reduce lighting energy include the following:
- The lighting system is designed to 0.65 W/ft^2, exceeding the energy code by 40%. Task lights were provided for the workstations but are rarely needed.
- The second-floor work areas are daylighted using solar tube skylights (tubular devices) to enhance the natural light from the windows. This work space also includes a custom light-emitting diode (LED) pendant chandelier and clerestory windows.
- Exterior offices adjacent to the daylighted areas have both exterior and interior windows, which allow more natural light into those work areas. Interior blinds reduce glare during winter when the angle of the sun is lower.
- Digital addressable daylight control systems modulate the electric lighting output in the work areas.

CMTA Headquarters Exterior

HVAC System

The building uses high-efficiency, geothermal WSHP units with an installed capacity of 600 ft²/ton and a geothermal well field capacity of 780 ft²/ton. The well field consists of 12 vertical bores, which are each 400 ft deep. The system includes dual refrigerant compressors for all heat pumps three tons or larger, which increases efficiency to roughly 23 SEER during part-load conditions, and a distributive water pumping system to recirculate water through the geothermal loop. Separate water pumps are sized to each heat pump and only operate when that heat pump compressor is on. A single rooftop unit that is equipped with a supply fan, an exhaust fan, an energy recovery wheel, and variable fan drives serves as a DOAS. A DCV system supplements the DOAS and modulates the outdoor air coming into the building in response to actual space conditions.

Energy Performance

CMTA tracks and benchmarks energy performance—the first year's energy consumption was 13.6 kBtu/ft²·yr, which earned the all-electric building an ENERGY STAR rating of 100.

An ethernet-based digital electric metering system is used to measure the energy consumed by the HVAC, lighting, and plug loads. Almost one-third of the annual energy use is consumed by plug loads. While this seems large, high-performance systems reduce the energy use of the HVAC and lighting systems, thus increasing the percentage of plug loads as compared to the total energy.

Additional information on the CMTA headquarters building and systems along with what was learned during the design, construction, and early operation of the building can be found in the Winter 2011 issue of *High Performing Buildings* (Seibert 2011).

CMTA First-Floor Conference Room

Second-Floor Daylighted Work Space

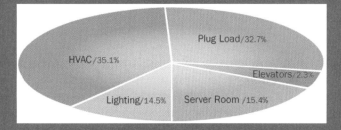

Metered Energy Use Breakdown

Photographs reprinted with permission of High Performing Buildings

Budget Sharing

One oft-heard but fundamentally unnecessary question is "whose budget pays for improved energy efficiency?" The answer should always be "the owner's budget!" When a team commits itself to delivering low-energy, holistic solutions, it is virtually impossible to discretize for the accountants how much energy efficiency each trade or discipline "purchased" on behalf of the project through its respective design decisions. A classic example is the cost of shading: there are increased structural and façade costs, but these may be offset by reduced capital cost for window glazing and air conditioning. These trade-offs are absolutely necessary to explore in consideration of the particular goals and context of the building. So long as the overall building construction budget remains consistent with the Owner's Project Requirements (OPR), it doesn't matter where the money was spent if the whole building performs.

What this tells us, however, is that discipline-based construction budget allocations might be inappropriate for the integrated design paradigm and should be reviewed early in the project. Similarly, it might be argued that traditional fee percentages may also be unintentionally preventing the disciplines most capable of proposing and proving energy-reduction techniques from applying their analytical technologies and abilities to the solutions.

Lastly, the EUI "budget" itself must also be equitably shared. The building envelope does not consume energy but significantly affects the energy use of mechanical and lighting systems. Legislation and ingenuity have brought us to the point at which most electrical, mechanical, and lighting equipment has been optimized for the current state of technology. Therefore, it is important for design teams to carefully review the relative proportion of energy use by discretionary design choice and collectively attack those portions of the pie chart that represent the greediest users. A classic example is the use of all-glass façades with the expectation that highly efficient HVAC systems will somehow accommodate the egregious gesture; thankfully, energy codes are now biased to avoid this practice. Another more subtle example is the issue of plug loads in highly efficient buildings. As the intentional reduction of lighting and mechanical energy use is applied, the plug loads grow in a relative manner to upwards of 50%. This should immediately tell all parties that plug loads need to be addressed, either with automatic shutdown controls or with substantial reduction in required, desired, or assumed load on the part of the owner and design team. If the team knows that it is accountable for sharing the responsibility for the end energy-use burden, it sets the tone for sharing the energy-savings burden as well.

Investment Financial Analysis

Many of the examples thus far have discussed trade-offs made by the design team to reduce the total building energy use. In order to confirm that each decision contributes to affordable energy savings, energy modeling can be coupled with a series of financial analyses to show which ECMs give "the biggest bang for the buck." The three most typical tools include the following:

- *Life-Cycle Cost Analysis (LCCA)* is a calculation method that adds first cost to 20–25 years of annual energy and maintenance costs, inclusive of equipment replacement costs and an estimate on inflation. The option that has the lowest life-cycle cost is usually chosen if the budget allows. LCCA is the financial tool most often used by institutional owners planning to hold and operate the building through a few generations of equipment technology.
- *Simple Payback Period* is a calculation method that divides first cost by the annual energy savings to determine how long it will take to break even on the investment. The simple payback is most often used by developers looking to recoup costs before divesting of a property or by long-term building owners with limited funding for retrofits.
- *Return on Investment (ROI)* is a calculation that takes the ratio of the energy savings over a predefined number of years minus the first costs divided by the first costs. It essentially answers the question "what is my rate of return on the investment?" and allows a somewhat parallel comparison to the rate of return used in the financial markets. The ROI

method is usually used by wealth-holding clients comparing relative opportunity costs when looking to invest in stable profit growth. In downturn economies burdened with the ever-rising cost of energy throughout the world, some financial institutions have begun to provide financing for energy-efficiency upgrades based on projected ROI through vehicles based on ROI calculations.

It is important in all of these financial comparisons that the team agree on which inflation and depreciation rates are appropriate for use.

Building Configuration and Floor Area Minimization

In the area of building design, the first item to address is the built area. For first-cost reasons, there is obviously a drive toward the minimization of built square footage, and the entire team should review the actual requested occupancies to determine if space can be shared as flexible space between uses otherwise listed separately. For instance, shared conference spaces or lounge spaces can reduce the redundancy of built space while also encouraging interdepartmental synergy. Another area often under scrutiny for cost savings (both first cost and operating costs) is the transient gross square footage associated with circulation space and lobbies. It is recommended that the team use space-planning exercises to review if there are ways for these types of spaces to be reduced in size through merging with other functions or to be limited in scope and controllability with regards to expenditure of energy under low-occupancy conditions.

The second major item for the team to address is the architectural configuration of the building. Façade square footage represents a source of conductive heat loss or heat gain as the OA temperatures fluctuate; therefore, the larger the amount of façade area, the greater this impact. Additionally, most façades for office buildings contain windows for the benefit of the occupants. Glazing is a poorer insulator than most opaque constructions and should be reviewed with regard to its placement and size. Generally speaking, daylighting and natural ventilation are possible within about 25 ft of a façade, a value that may govern the depth of footprints aspiring to greater connectivity to the outdoors.

Beyond the impact on the interior floor plate, the shape of the building also informs where and how the building self-shades and begins to inform where glazing can be most effectively placed. Generally speaking, in the northern hemisphere, glazing that points toward the north captures sky-reflected daylight with minimal solar heat content, making it the ideal source of even light. Eastern and western glazing is impacted by low-angle sun throughout the year, which can cause glare and thermal discomfort if not mitigated properly. Lastly, in the northern hemisphere, southern façades with glazing benefit from overhangs to reduce solar load during the summer season.

Safety Factors and Diversity Factors

It is quite important for all members of the design team to openly reveal their safety factors so that systems are not oversized. The judicious application of diversity factors based on how normal buildings operate is important to the tight control of rightsizing the equipment for optimum efficiency. A classic example is the plug-load allowance requested by the owner. The owner knows what the nameplates are—they can be up to four times higher than normal actual operating levels (ASHRAE 2009b, Table 8). The HVAC designer accepts that load and then applies a factor of +20% to account for future expansion and then, as per the code, is allowed to size equipment for an additional 20%–30% for morning warm-up and boost. Then the fans are all sized for an extra 10% for air leakage, and the electrical engineer takes the mechanical loads and adds an extra 15% for unforeseen additional load or for effectively following the *National Electrical Code* (NFPA 2011) by taking everything at face value simultaneously. All told, one can find transformers sized more than three times larger than the largest load ever likely to be experienced. The result of this drastic oversizing is that some equipment may be operating in inefficient ranges, distribution flows may become unstable at low turndown rates, and excessive material has been installed as compared to what was actually needed. It is strongly recommended that a map

The Terry Thomas—A Case Study

The building team for The Terry Thomas, a 40,000 ft^2 Class A commercial office building in Seattle, used an advanced approach to energy-efficient design with the use of natural ventilation and daylighting. Extensive thermal and energy modeling was used to determine the most effective strategies and balance the trade-offs needed to design a building with passive cooling, natural ventilation, and daylighting while maintaining a comfortable work environment for the occupants. The LEED Gold-CS, LEED Platinum-CI building was measured as 42% more efficient than a baseline building designed to ASHRAE/IESNA Standard 90.1-2004. Moreover, through the integrated design process, the project achieved significant energy efficiency within the initial construction cost budget.

The building is designed as a "square doughnut" form that enhances the natural ventilation and daylighting while providing communal space for the tenants via the center courtyard. The passive cooling strategies employed are complemented by automated louvers and a combination of dynamic and fixed external shading.

Shading and Daylighting

The optimal designs required that two goals be balanced: the need to remove solar heat and glare that might hinder natural ventilation and the need to provide adequate daylight to interior spaces. Eliminating solar gain also allows for the elimination of the mechanical cooling system while still maintaining thermal comfort. Effective external shading was the key to controlling solar gain and thus reducing cooling peak load. The project optimized the shading strategy for each façade at each floor:

- Fixed exterior tinted-glass shades ("sunglasses"), angled to provide the required shading and optimum levels of daylight, are used on the east and west external façades.
- Dynamic external venetian blinds that are automatically adjusted based on light level and sun angle are used on the northeast, south, and some east façades.
- The building shape and courtyard allow for uniform daylighting in each open office space. The high ceilings and narrow floor plates allow natural light to penetrate the interior.
- Daylighting is further enhanced with high internal reflectance, exposed ceilings with no ductwork, and low open office partitions.

Building Data	
Steel deck roof with gypsum sheathing	Overall R-value = R-30 Reflectivity = 95
Metal-framed walls with gypsum sheathing, batt insulation, and corrugated metal cladding	Overall R-value = R-21 Glazing = 45%
Windows	U-factor = 0.4 SHGC = 0.36 VT = 68%
Sunshades	SHGC = 0.52
EUI	37 kBtu/ft^2
Annual cost index	$0.53/ft^2·yr
Total construction cost	$10.2 million
Cost per square foot	$160/ft^2

Natural Ventilation

The building uses passive cooling and natural ventilation, and heating is provided for the building perimeter by radiant hydronic baseboards. Internal heat gains are reduced by the use of less artificial light as a result of the maximized use of natural light. The following characteristics are the key components of the natural ventilation system.

- The internal courtyard allows for cross-ventilation, which is enhanced by the narrow floor plate and open office plan. Other features that help to optimize the natural ventilation system include the open ceilings and castellated beams.
- Motorized louvers are located near the ceiling on each floor and are used in conjunction with occupant-controlled operable windows along the building perimeter.
- The passive cooling design and use of natural ventilation allowed for the elimination of the mechanical cooling system and reduced the HVAC equipment cost. Louvers are controlled with CO_2 sensors. This use of this DCV strategy contributes to the reduction of unnecessary outdoor ventilation air and the accompanying heating energy use.

Additional information on The Terry Thomas and the extensive modeling that was used in the design process can be found in the Summer 2010 issue of *High Performing* Buildings (Sethi and Marseille 2010).

Second Floor Office on the South Façade
with Castellated Beams*

Exterior Tinted-Glass Shades
Used on the East and West Façades[†]

External Venetian Blinds*

Internal Courtyard*

*Photographs reprinted with permission of Chris Meek and Kevin Van Den Wymelenberg

[†]Photograph reprinted with permission of High Performing Buildings

of all safety factors and diversity assumptions is clearly laid out in a transparent way so that the whole team can make judgments together with regard to the assumptions.

Diversity factors are very different from safety factors. The former are meant to deal with known uncertainties or fluctuations based on professional judgment and industry practice, while the latter are meant to deal with unknown uncertainties in the operations in the future. For instance, in HVAC and electrical design, it is quite common to find the following diversity factors applied:

- Solar diversity, embedded in most computer modeling software in calculating energy use and total peak load
- Diversity assumptions about occupant attendance (i.e., a 75% diversity factor on office occupancies is quite common)
- Diversity assumptions about computer use with links to occupancy diversity and integrated with assumptions regards computer activity kilowatt use versus sleep mode, given occupant presence at the desk
- Diversity assumptions with regards to likely simultaneity of peak airflows or peak water flows occurring on a single system—this is often used to downsize system capacity

It is important to note that diversity factors are independent of schedules and as such must be reviewed with the schedules to ensure that the appropriate level of fluctuation is accounted for only once (especially when the schedule is a percent of load schedule). It is crucial for the entire team to agree to the diversity factors, as using them to downsize equipment for energy efficiency might run the risk of reduced capacity on peak days. It is necessary to project the extreme internal conditions arising with these peak conditions and get buy-in from the owner for exceedance over a known number of hours in a typical meteorological year.

Schedules of Occupancy, Use, and Utility Rates

It is essential that the team understand the schedules related to utility rates, especially any embedded demand charges and on/off/hi/low/seasonal peak period definitions local to the site and its service utility. This is because the prevailing benchmarks for energy savings in ASHRAE/IES Standard 90.1 (ASHRAE 2010b) and most energy codes are based on annual cost, not absolute energy savings. Most importantly, the owner pays for the demand and consumption charges. This means that discretionary decisions by the team to avoid onerous demand charges through load shifting may be appropriate when looking to reduce annual operating expenditures.

It is important for the team to map out the anticipated schedules of use and occupancy for each area of the building. This is information that is crucial to the energy modeling and can greatly affect the outcomes with regard to estimated energy savings over a known benchmark or LCCA. It is important to note that most energy models run the same schedule week after week, so schedules not only should be configured to cover typical weeks but also should be changed to account for any known long periods of building closure.

It is important also to look at how the relative schedules interact with each other. For instance, the following assumptions and techniques are often used in energy modeling in the office context:

- *Lighting schedules* in areas with occupancy sensors are often arranged to have 100% light for any nonzero level of occupancy.
- *Small power loads representing office equipment* are often arranged to have a percentage load to match the percentage of occupants, except for a 10%–15% nighttime parasitic load for equipment in sleep mode or that is not automatically turned off.
- *Server room load and HVAC equipment cycling*—it is important to note what proportion of the server room server load is related to occupants and their immediate usage as well as how much is due to continuous processing.

- *Kitchen/lounge/vending equipment*—this equipment tends to have high usage during the midday hours. It is important to examine the schedule of operation of such equipment as microwaves and heating plates as well as the operation of any exhaust fan associated with odor control in the area. Refrigerated vending machines and refrigerator/freezers tend to operate throughout the day to cycle to maintain internal conditions. It should be noted that newer vending machines have occupancy sensors that trip on the display lighting for the patron but then revert to a darkened mode.
- *Conference facilities*—engineers must make a judgment in energy modeling as to whether or not the conference facilities should be modeled as independent zones with an occupancy schedule. The reason for this is that there are different ways of thought about these areas. Some argue that the people in the conference room are the same as the people who would have been in the adjacent offices, so one should just have a large zone inclusive of the conference and office areas, with the appropriate diversity applied during load calculations but not during energy modeling. Other engineers argue that set-aside conference rooms do have an increased instantaneous occupancy that can drive up overall ventilation rates in systems serving both conference rooms and offices, so the true model should reflect the increased load. The problem with the latter approach is that the occupancy of the conference rooms will begin to form a base load every hour of every weekday for every week in the energy model. It is necessary to review the impact of this purely modeling decision on the projected energy use of the whole building, especially if there are a large proportion of conference facilities.

The last item to bear in mind regarding scheduling is whether a standardized schedule will be imposed on the energy model through regulatory requirements. For compliance modeling, in particular, some codes such as the *California Energy Code* (CBSC 2011) require that prescribed schedules are used instead of schedules grounded in realistic reviews of assumptions. It is important for the entire team to be aware of any such constraints ahead of time.

Redundant and Standby Capacity Sizing Protocols

It is recommended that a thorough discussion with the owner about redundancy be conducted early in the design process. In particular, how redundancy is achieved and whether it is necessary should be discussed. Redundancy is the creation of spare capacity such that a single piece of equipment can be down for maintenance and the rest of the system can operate at some level. It is usually the case that percentage of capacity is used to define redundancy.

For instance, in a system in which two pumps are each sized at 60% of total load, the facility can lose one pump and still have at least 60% capacity. In a real-world installation, it is necessary to ensure that the selection of that pump has the optimum efficiency for normal usage instead of an artificially high peak design flow.

If 100% capacity is desired at all times, some engineers prefer to use standby equipment—a whole spare unit capable of running when any of the normal run equipment is not functional. In this case, it might be appropriate to have three pumps, each sized for 50% capacity, available in the system—this would be a two run/one standby approach, and the pumps would generally be rotated in their operation in order to equalize runtime.

It is important in the energy modeling context that the appropriate horsepower associated with the actual 100% of capacity be used, lest the project be overburdened with a model of energy use that would not occur in real life, even when the model includes a variable-flow control.

Façade Zone Optimization

One key area of focus for the multidisciplinary team is the façade zone, the nexus of many design desires. The façade must perform multiple functions simultaneously to create an adequate space for the occupants inside. The drivers for this include the following:

1. Presenting the public "face" of the building
2. Protecting the indoors from wind and rain

3. Sealing the building envelope to contain conditioned space
4. Creating a security barrier
5. Creating a healthy and comfortable environment for occupants (thermal, acoustic, visual, and IAQ)
6. Mitigating solar heat to be removed by a comfort-conditioning system
7. Tempering the influence of cold and hot temperatures outdoors
8. Controlling noise break-in from the outdoors
9. Manipulating or enhancing daylight for useful purpose
10. Providing views to the outdoors to improve occupant satisfaction
11. Moving air in and out of the building (operable elements or louvers)
12. Maximizing usable/leasable square footage indoors

The design of façades and the perimeter zone immediately behind them is particularly difficult precisely because there are competing priorities that must all be resolved in order to deliver a functioning building. Having the entire team acknowledge the full range of needs is the first step, and then they can look for integrated solutions together given a limited façade budget.

In addition to the competing design needs, current technology in façade materials also creates "forced marriages" that must also be managed. For instance, very transparent glazing also usually comes with a high solar heat gain—the U-factor, SHGC, and visible transmittance (VT) come directly out of glazing selection. Table 3-5 documents some of the key points of contention that have impacts on energy efficiency of the perimeter zones and relates them to the drivers detailed in the list above (items 1–12 are referenced as drivers and/or secondary impacts in the table).

Table 3-5 Guidance for Improving Energy Efficiency in Perimeter Zones[a]

Glass Body-Tinted Color and Allowable Outward Reflectivity		
Drivers	(1) Public face, (10) Views to outdoors	High reflection and dark tint lower solar heat transfer AND lower VT.
Secondary Impacts	(6) Solar heat, (7) Cold/hot influences, (9) Daylighting	
Glass Unit Configuration (Insulated Glazing Unit, Laminated, Tempered?)		
Drivers	(1) Public face, (4) Security barrier, (5) Health/comfort (thermal, acoustic)	Laminated and tempered elements in the outer panes are usually required in ballistic or blast-proof designs. Insulated glazing units (IGUs) are the energy-efficiency norm in most climate zones to improve U-factor and are beneficial in increasing the Sound Transmission Class noise transmission value of the window.
Secondary Impacts	(6) Solar heat, (7) Cold/hot influences, (8) Noise break-in	
Use of Low-e Coatings		
Drivers	(7) Cold/hot influences; (5) Health/comfort	Low-e coatings are usually placed on either surface 2 or 3 in IGUs in order to retain heat in winter or reject solar heat in summer. They reduce U-factors but can sometimes slightly effect VT.
Secondary Impacts	(10) Views to outdoors; (9) Daylighting	
Use of Spectrally Selective Interlayers, Coatings, or Glass		
Drivers	(6) Solar heat; (5) Health/comfort (thermal)	These improve SHGC directly while trying to preserve VT as high as possible.
Secondary Impacts	(9) Daylighting	

[a] The numbers in the lists of drivers and secondary impacts correspond to the numbered items in the "Façade Zone Optimization" section of this Guide.

Table 3-5 Guidance for Improving Energy Efficiency in Perimeter Zones[a] (Continued)

	Use of Spandrel Glass, Especially at Ceiling Voids	
Drivers	(6) Solar heat; (7) Cold/hot influences	This places insulation behind glass in order to improve the local R-factor and reduce HVAC loads. It tends to look like a more "solid" or "colored" version of the glass, which is noticeable from outdoors.
Secondary Impacts	(1) Public face; (10) Views to outdoors	

	Use of Operable Windows/Louvers Controlled by Occupants	
Drivers	(11) Air movement; (5) Health/comfort (thermal, IAQ)	Windows are often seen as an amenity that have the potential to reduce energy use if occupants are trained to open them during temperate weather. Care must be taken to address the Secondary impact issues with the façade detailer and the owner, and mock-up testing is recommended to validate quality of seals.
Secondary Impacts	(1) Public face; (3) Building seal; (4) Security barrier; (2) Wind and rain	

	Use of External Shading	
Drivers	(6) Solar heat; (1) Public face (if feature); (5) Health/comfort (thermal)	External shading comes in all shapes and sizes, but the following have the most success in the northern hemisphere—overhangs at south, fins at north if needed for early morning or late afternoon solar angle, and parallel perforated or movable shading at east and west to allow views once the sun has moved.
Secondary Impacts	(1) Public face; (10) Views to outdoors; (9) Daylighting	

	Use of Internal Blinds	
Drivers	(5) Comfort (visual)	Internal blinds are not ECMs since they still allow the heat to enter the occupied space where the air conditioning must absorb it. Internal blinds are measures to avoid glare and to mitigate direct solar heat gain affecting occupants. Heavier internal blinds during high solar intensity do help to confine heat to the space behind the blind, and detailing should ensure that there is a gap at the top to allow overheated air to rise to the ceiling. Heavier internal blinds pulled during the winter can act to reduce the radiative cooling experienced when sitting near a large piece of glass.
Secondary Impacts	(1) Public face; (10) View to outdoors; (9) Daylighting; (6) Solar heat; (7) Cold/hot influences	

	Area of Glazing by Orientation	
Drivers	(1) Public face	In the northern hemisphere, it is usually recommended that the amount of glazing in east- and west-facing façades be limited, as the solar angles are low and the sun's intensity is hard to mitigate. Large areas of north-facing glass can be used for even-intensity daylighting with a lessened risk of solar heat gain as compared to southern glass, which should be protected through external shading devices.
Secondary Impacts	(6) Solar heat; (10) Views to outdoors; (9) Daylighting; (8) Noise break-in	

	Use of Curtain Wall Systems versus Window Wall Systems versus Punched or Ribbon Openings	
Drivers	(1) Public face; Envelope seal	Cost is usually the primary driver for this issue; however, envelope sealing and thermal bridging detailing should be carefully addressed in all non-curtain wall systems. For curtain wall systems, noise-flanking blocks and fire/smoke blocks at the floor line require careful attention.
Secondary Impacts	(6) Solar heat; (3) Security barrier; (10) Views to outdoors; (9) Daylighting; (2) Wind and rain; (8) Noise break-in	

[a] The numbers in the lists of drivers and secondary impacts correspond to the numbered items in the "Façade Zone Optimization" section of this Guide.

Table 3-5   Guidance for Improving Energy Efficiency in Perimeter Zones[a] *(Continued)*

	Use of Double-Façade System with Interpane Blinds	
Drivers	(1) Public face	Energy savings from double façades tend to pay for themselves only in heating-dominated climates (climate zones 4A, 4C, and 5–8). In winter, the cavity can be sealed to act as a large insulated buffer zone and to absorb heat in the dark surfaces of interpane blinds. In summer, the cavity is opened, and the solar-heated blinds cause a convective updraft that continues to remove heat from the cavity, allowing the inner pane of the inner IGU to react only with the indoors.
Secondary Impacts	(6) Solar heat; (3) Security barrier; (9) Daylighting; (11) Air movement; (8) Noise break-in; (5) Health/comfort (thermal)	

	Strategy to Decouple Daylight Glazing from Vision Glazing	
Drivers	(1) Public face	Vision glass requires a clarity of perception that is not necessarily required for daylighting glass, which is meant to diffuse light and allow it to penetrate into the space. Some options for daylighting glass include fritted glass, spectrally selective glass, prismatic etching, diffusing elements, and light shelves.
Secondary Impacts	(6) Solar heat; (9) Daylighting; (5) Health/comfort (visual)	

	Strategy to Place Heating Elements within, at, or near Façade Construction	
Drivers	(7) Cold/hot influences; (5) Health/comfort (thermal)	Perimeter heating convective or "radiator" elements are beneficial because they deal with the heat loss in the façade while allowing the room supply airstream to react just to room temperature, which results in a lower risk of overheating. Some attention to controls is necessary to avoid significant simultaneous heating and cooling.
Secondary Impacts	Leasable space	

	Strategy to Place Cooling Air Discharges within, at, or near Façade Construction	
Drivers	(7) Cold/hot influences; (5) Health/comfort (thermal)	Similar to the heating elements above, some designers put supply diffusers at the floor to mitigate conductive and solar heat buildup in the glass and put a return diffuser at the ceiling in order to remove heat buildup. This can affect the intended desk position or space planning if not coordinated in advance. Unlike the heat loss situation, solar heat gain has directivity arising from the sun angle and is not confined just to the inner face of the façade.
Secondary Impacts	Leasable space	

	Thermal Bridging in Glazing and at Window-Wall Junctures	
Drivers	(7) Cold/hot influences; (3) Building seal	In many climates, thermal bridging in IGUs is standard to prevent the metal frames from acting as heat conductors. In cold climates, poor thermal breaks can lead not only to surface condensation but also occasionally to frost, which can cause mold or mildew problems when it melts. Even with thermally broken glazing, there is significant vulnerability at the interface between the window and the wall surrounding it. Continuous lines of insulation are required to complete the building envelope, and good seals are necessary to minimize infiltration.
Secondary Impacts	Leasable space	

	Condensation	
Drivers	(5) Health/comfort (IAQ)	See above for thermal breaks in window frames. To avoid condensation on the internal surface of glazing, review the selection of the glass U-factor to determine the likely internal surface temperature of the glazing on the design day versus internal dew-point temperature. If condensation on the window is a risk, increasing air velocity is recommended along with possibly providing a local duct heater on the supply diffuser to blow against the glass in response to a surface temperature feedback sensor.
Secondary Impacts	N/A	In opaque wall constructions, careful detailing is necessary to reduce the effects of condensation within the wall void. Condensation can lead to mold growth, which may have irritating, allergenic, asthmagenic, or toxic effects.

[a] The numbers in the lists of drivers and secondary impacts correspond to the numbered items in the "Façade Zone Optimization" section of this Guide.

Table 3-5 Guidance for Improving Energy Efficiency in Perimeter Zones[a] (Continued)

Downdraft		
Drivers	(5) Health/Comfort (thermal)	Glazing including IGUs is a poor insulator as compared to insulated opaque construction. As noted immediately above, the internal surface temperature of the glazed unit must be calculated in order to examine not only condensation but also the risk of creating localized cooled air arising from touching the surface in a conductive and convective heat loss from the air. This cooled air is more dense than the surrounding room air and tends to fall by gravity. Particularly with tall glazed surfaces, the cumulative effect of falling cooled air can create an unacceptably high velocity of the downdraft, which may cause discomfort. Intermediate mullion breaks or heating elements or increased localized air movement from heated room supply air blowing against the glass can help to mitigate these effects.
Secondary Impacts	N/A	
Selection of Room HVAC Equipment to Deal with Large Fluctuations		
Drivers	Comfort (thermal)	It is important to acknowledge the wider fluctuations that exist in perimeter zones as compared to internal zones. Please see the Building Zoning section later in this chapter. The radiation effect of cold glazing surfaces can also affect occupant thermal comfort.
Secondary Impacts	N/A	

[a] The numbers in the lists of drivers and secondary impacts correspond to the numbered items in the "Façade Zone Optimization" section of this Guide.

Plug Load Reduction

It is essential at the start of the project that the entire design team examine all assumptions related to occupant-affiliated electrical loads, or plug loads. Some plug-load equipment may be selected by different disciplines within the design team and other items may be selected outside the design process. If possible, the building occupant should also be involved in identifying plug-load equipment and controls.

Plug-load equipment gives off waste heat, which can increase cooling. Controlling these loads may also reduce required cooling equipment sizes. In many office buildings, developers and owners set arbitrary plug-load equipment power densities in units of W/ft^2 as their required available capacity in order to make the space marketable. But more is not always better when it comes to plug loads, especially if the air conditioning is sized to suit. Oversized HVAC equipment trying to work at low part load can result in excessive energy use and temperature instability in the spaces. Identifying the actual expected plug-load equipment power needs can allow closer sizing of HVAC equipment, reducing cost and improving energy performance.

Electric Lighting Load Reductions

Lighting loads, similar to plug loads, have a multiplicative impact on overall building energy use and should be addressed with a similar set of occupancy-tracking steps as noted above.

- Lighting loads can first be reduced by reviewing lamp, fixture, and ballast efficiency along with the lighting design to determine if lower wattage per square foot of installed lighting density can be used while still achieving the necessary lighting levels.
- Open office spaces should be located on the north and south sides of the building to maximize the daylighting potential.
- There is an industry-wide practice of using occupancy sensors and scheduling to control lighting during normal occupancy. It should be noted that most jurisdictions also allow the application of occupancy-sensor controls on egress lighting, often called *24-hour* or *night lighting*, to further reduce electricity associated with lighting an unoccupied building.
- All exterior lighting should be reviewed for its necessity for security or appearance reasons, and the lowest exterior LZ should be used with exterior lighting systems.

- While ASHRAE/IES Standard 90.1 (ASHRAE 2010b) allows for "exempt lighting equipment," this lighting is only exempt from the LPD requirements, and these exempt lights still produce heat that must be removed from the building. Avoid the temptation to use exempt lighting equipment if the design can be accomplished within the lighting power budget.

Conditions to Promote Health and Comfort

Near the start of a project, the design team should engage in a serious discussion of occupant health and comfort in order to define the criteria to be applied to the project. This discussion should cover the following aspects.

Indoor Air Quality (IAQ)

IAQ is an essential aspect to be discussed at the start of the project in order to reduce health risks for future occupants. Good IAQ must always be a priority when considering all design decisions and must not be adversely affected when striving for energy reduction. ASHRAE Standard 62.1 (ASHRAE 2010a) provides nominal guidance on minimum requirements for ventilation, but it should be acknowledged that IAQ encompasses far more than just ventilation.

As noted in *Indoor Air Quality Guide: Best Practices for Design, Construction, and Commissioning* (ASHRE 2009c), the following key objectives must inform the work of the IPD team:

- Manage the design and construction process to achieve good IAQ
- Control moisture in building assemblies
- Limit entry of outdoor contaminants
- Control moisture and contaminants related to mechanical systems
- Limit contaminants from indoor sources
- Capture and exhaust contaminants from building equipment and activities
- Reduce contaminant concentrations through ventilation, filtration, and air cleaning
- Apply more advanced ventilation approaches

Care and judgment must be applied at all stages of the process to ensure a healthy environment for indoor occupants. For more information on IAQ, refer to *Indoor Air Quality Guide* and its specific recommendations.

Thermal Comfort

The design team discussion should begin by determining the normal activity level of the occupants in each main zone (as compared against *ASHRAE Handbook—Fundamentals* [ASHRAE 2009b]) and the range of mandatory or voluntary dress code will be allowed in the building. The conversation should then cover the main concepts of dry-bulb temperature, relative humidity, operative/"effective comfort" temperature, and ASHRAE Standard 55, the comfort standard (ASHRAE 2010c). What range of operative temperatures will be considered "comfortable" for the various spaces and whether this collection of comfort temperatures may vary in response to seasonal changes should be agreed upon. If possible, surveying existing tenant staff members at a similar facility to benchmark attitudes regarding thermal comfort can be beneficial. Once the criteria are set, then it is possible to produce a simplified overview of the relative effects of convective and radiative heat transfer, the reduction of evaporative and respiratory heat rejection that occurs with increased humidity, and the use of increased air movement to improve convective heat transfer in a "wind-on-skin" compensation for higher setpoints for room temperature.

Clothing plays a major part in reducing energy usage. Dressing in layers allows an individual to respond to a changing interior environment. For example, in zones facing the equator (south in the northern hemisphere), the low sun angle significantly increases the penetration of direct sunlight into the building. The perceived temperature can change significantly throughout the day due to direct sunlight falling on occupants. Similarly, as noted in Standard 55, radiant temperature also can play a significant part in comfort, particularly in the area of asymmetry. Table 5.2.4.1 of Standard 55 provides guidance on surface temperature differentials to be maintained to avoid asymmetry.

Per Standard 55, typical considerations for thermal comfort include

- metabolic rate,
- clothing insulation,
- air temperature,
- radiant temperature,
- air speed, and
- humidity.

Lastly, all parties should consider allowing a wide dead band for occupied-mode setpoints as a measure to reduce energy use (as compared to the minimum dead-band range stated in energy codes); however, these expanded temperature ranges should not be so extreme as to compromise the productivity in the work space.

Visual Comfort

Lighting, both daylight and electric light if designed and integrated properly, will minimize visual comfort issues in the space.

Electric lighting levels should be designed to meet IES recommended light levels (IES 2011). Providing light levels that are too high or too low will cause eye strain and loss of productivity.

Direct sun penetration should be minimized in work areas because the light level in the direct sun can reach 1000 or more footcandles (versus 30 to 50 footcandles from electric lighting). This high contrast ratio will cause discomfort issues. Using light shelves on the south side of the building will minimize the direct sun penetration for workers near windows while allowing daylight to penetrate deep into the building.

Worker orientation to windows is also very important in minimizing discomfort issues. Computer screens should never be positioned facing windows (with workers' backs to the windows) or facing directly away from the window (with workers facing out the windows). Both of these situations produce very high contrast ratios, which cause eye strain. Locate the computer screen and worker facing perpendicular to the window to minimize worker discomfort.

Further recommendations for lighting visual comfort can be found in the tenth edition of IES's *The Lighting Handbook*, specifically in Chapter 2, "Vision: Eye and Brain," and Chapter 4, "Perception and Performance" (IES 2011).

Acoustic Comfort

The design team discussions should cover ambient noise criteria for each space, acoustic privacy, occupant-created background noise, and speech intelligibility. Because this Guide covers primarily design decisions related to energy efficiency, the primary topic of focus regarding acoustic comfort is noise criteria, as lower noise in the space may require the application of duct silencers, which tend to increase the friction experienced by the system and thus increase energy use. Noise criteria in the space may also require acoustic ceilings and carpets, which tend to prevent the optimal activation of thermal mass for human comfort.

The second critical topic uniquely related to acoustics and energy in office buildings is the question of open-plan office spaces. Many designs incorporate private offices at the perimeter of the building, which confines the benefits of a natural ventilation scheme to only those few individuals in the offices (many of whom are managers who may not be present as often as lower-ranked personnel). Placing open-plan spaces at the perimeter for possible energy efficiency benefits should be weighed against corporate hierarchies. The predominant use of open-plan offices spaces at the interior of the building can allow for multiple low-energy cooling techniques to be applied, such as displacement or underfloor air distribution, radiant ceilings or floors, and chilled beams. The reason these low-energy systems are appropriate is that the heat loads are relatively constant and are at low density as compared to areas experiencing solar heat. The design team should acknowledge that multiple studies show that most occupants dislike the lack of acoustic privacy that arises in open-plan office spaces, regardless of cubicle partition height. Some consideration of artificial, low-energy noise-masking white noise or pink noise speakers should be considered, especially for low-energy HVAC systems, as their characteristically low velocities do not create that same level of white noise as overhead high-airflow systems.

With regard to the speech privacy and intelligible noise issue, the design team should consider whether there are other space-planning techniques to resolve it, such as small 5×5 ft "telephone booths" that allow individuals to make private calls or that can accommodate a couple of people gathering around a phone for a teleconference to avoid irritating their neighbors.

Building Zoning

During its space-planning exercise, the design team should spend sufficient time in discussion to understand how the building will be zoned for HVAC and lighting, as this will effect the possibility for future design applications to achieve energy efficiency.

Much discussion has already been made of the benefits of understanding solar heat gain in the perimeter zone. To summarize, perimeter glazed zones can have fluctuating availability of daylighting, seasonal capacity to apply natural ventilation capacity, and extreme reactions to seasonal or diurnal changes in weather. The HVAC and lighting zoning of these spaces must be responsive through the entire range of extreme conditions. Perimeter private-office spaces should be clustered on HVAC zones based on common orientation and should take into account the political aspects related to the number of people who will be arguing over the setpoint temperature control, as usually only one office can be master but all offices will get similar quantities of cooling air. Corner offices with glazing on both façades should always be their own HVAC zones. Lastly, perimeter zones should be ducted and piped to allow for perimeter heating and nighttime building pressurization/setup and setbacks to operate while turning off all centralized air conditioning associated with the interior zones. Internal zones consisting of office occupancies usually are fairly stable with regard to heat load and occupancy and should be independently zoned to take advantage of these facts while using lower-energy systems. If nighttime shutdowns of plug-load equipment and lighting are instituted as noted above, most interior zones can suffice with shutting down the air handlers and doing a purge cycle of preconditioning in the morning.

High-occupancy-density zones, such as classrooms or conference rooms/lounges, should be zoned independently from areas having more stable and lower-intensity occupancy. High-density areas should have occupancy-based DCV in order to automatically reduce the amount of ventilation air provided to the space and thus the amount of OA cooled or heated to suit. It is recommended that the ventilation air in these zones be decoupled if possible from the overall cooling supply.

High-plug-load/heat-density spaces are usually computer rooms, electrical/information technology/audio-visual/security rooms, and server rooms in office buildings. These zones should be completely independent from the rest of the building, as this equipment tends to run 24 hours per day and requires cooling throughout that period. These spaces are often best served by a local recirculation unit with a cooling coil moving large amounts of cooling air. This approach tends to significantly reduce fan energy as compared to using remote or central systems. It should be carefully noted that if a situation arises after the fact in which a space formerly designed to reside on a centralized office-type system is packed with high-plug-load electronic equipment, there is a risk for significant inefficiencies, as this one overpacked zone will drive the central air handler's supply temperature to the minimum and the rest of the zones will experience significant reheating of the supply airstream to meet comfort conditions. In these cases, placing a local recirculation fan-coil unit in the room will help to absorb heat locally while still allowing ventilation air to come from the central system.

Despite the optimally energy-efficient zoning techniques espoused above, it is sometimes the case that zoning of HVAC and lighting systems needs to occur by department for back-charging purposes. The design team should discuss the relative first cost of submetering components with the expectation of annual energy savings versus the first-cost benefits of combining disparate zones. Alternatively, per-square-foot back charges can also be calculated based on anticipated proportion of annual energy use arising for each zone, based on energy modeling.

Controls

When striving to achieve significant energy reductions on the order of 50% savings toward net zero, the appropriate application of automatic controls is necessary, as occupants of the building are primarily and appropriately focussed on their day-to-day activities, not on behaving in an energy-conscious manner. A number of control strategies are discussed in Chapter 5; these will not be discussed here. This section is devoted to talking about multitrade integration of controls.

The following issues should be considered for integrated approaches in controls:

*   Shared information technology backbone and routing (if data security will allow it) to improve Web-based access to energy-use data.
*   Shared connection of occupancy sensors between HVAC and lighting controls.
*   Motorized blind control algorithms in response to anticipated solar heat gain and glare or to provide insulation in the heating season.
*   Facilities scheduling software interlinked into HVAC ventilation controls.
*   Server-room load-management controls as noted in the multidisciplinary "Plug Load Reduction" section earlier in this chapter.
*   Consolidated reporting of motor kilowatt-hours versus anticipated benchmarks.
*   Energy-use dashboards showing instantaneous energy use or monthly energy cost roll-up by zone to encourage departmental competition and behavioral change.
*   Overlays of plug-load monitoring, lighting, and HVAC controls to monitor relative energy use intensities by use and by zone throughout the day and year. This is useful in troubleshooting energy hogs.

Commissioning

The following multidisciplinary activities and the associated personnel should be considered for integrated approaches in traditional mechanical, electrical, and plumbing system Cx. Quality assurance, including Cx, is discussed in more detail in Chapter 5. Construction document specifications include requirements for Cx activities, such as participating in and documenting results, commissioning meetings, collaboration, and corrective actions.

*   Site-based Cx requires input from at least the following parties: the general contractor; the mechanical, electrical, controls, and test and balance (TAB) subcontractors; the CxA; the owner's representative; and the mechanical, electrical, and lighting designers.
*   Pre-functional test procedures usually require evaluation of motors and wiring by the electrical subcontractor and the manufacturer's representative and evaluation of component performance by the manufacturer's representative and the mechanical, TAB, and controls subcontractors. The CxA will generally sample to back-check the values reported in the pre-functional checklist results.
*   Functional tests involve the CxA and the controls and TAB subcontractors at a minimum.

Besides the usual tests of control sequences, it is also important to document that the building is ready from an IAQ point of view, as it is necessary to remove the construction-related odors and off-gassing chemicals from the air volume of the space prior to permanent

Energy-Use Dashboards at ASHRAE Headquarters

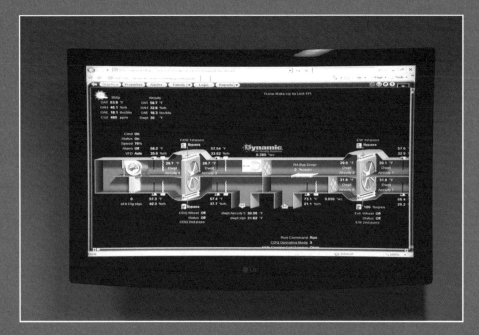

Makeup Air Unit Schematic
Source: Michael Lane

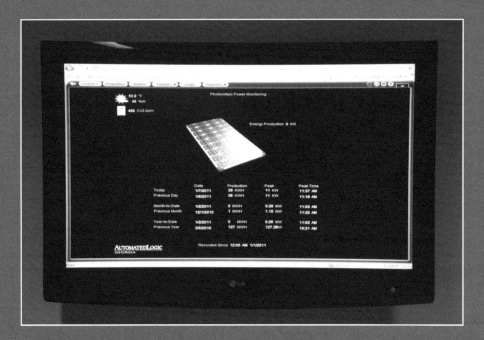

PV Power Monitoring
Source: Michael Lane

When ASHRAE renovated their headquarters building in 2008, one of the features that was added was real-time monitoring of the building subsystems, which is displayed continuously on screens in the building's lobby. The systems monitored include the makeup air unit and the PV power.

occupancy. This can be accomplished through physical testing in which concentrations of typical pollutants are measured and compared to health standards. It can also be proven (with agreement of all team members) through the careful documentation of preoccupancy purge procedures, which usually involve multiple hours of 100% ventilation air supply.

Charettes and Design Reviews

It is very desirable that design teams pursuing significant energy savings engage in the habit of early-phase design charettes involving all team members, followed by periodic design reviews. This process of holding one another accountable throughout the design process helps ensure that unintentionally myopic thinking on the part of any team member doesn't accidentally propagate into a vulnerability not caught until Cx. The entire team must understand the multidisciplinary multilateral agreements as noted herein, must acknowledge and support the achievement of stated energy-use goals, and must comb through the documents as they grow to ensure that the holistic system survives through detailing and value engineering processes.

Typically, kick-off charettes are convened by a named facilitator who sets the ground rules of the brainstorming session to encourage people to contribute and most importantly to listen. There should be agreed-upon time limits to each person's speaking length during brainstorming and time limits on the brainstorming period. All ideas are welcome and can be raised without dialogue or judgment during the brainstorming period. It is often useful to start with a brainstorming period related to project and team goals, followed by a discernment session that allows the "brain-dump" list to be ordered with prioritization for time and cost investment. This can then be followed by a "blue-sky" type of brainstorming related to energy-efficiency measures. During this brainstorming session, it is necessary to refrain from actually starting to design or else the value of the limited-time creative output from all team members may be diminished. There will be months to design using the great ideas thereafter.

Design reviews can benefit from reviewers that are both internal and external to the team. Internal reviewers are intimately aware of all of the step-by-step decisions that led to the current state. External reviewers provide a level of objectivity and can offer advice from past experience on similar challenges. The CxA's job is to review the content for commissionability and minimized energy use. Again, a facilitator may be necessary to ensure that all reviewers have time to speak without their suggestions being immediately contested by those parties with an investment in the status quo of the current design or who are biased for whatever reason. It is often beneficial to capture in writing all of the comments in an objective manner so that they can be respectfully addressed in sequence and so that a resolution on change of or continuance of design direction can be achieved and shared by the entire team.

USE OF ENERGY MODELING AS DESIGN GUIDANCE

Energy modeling is a powerful design tool to review the relative energy savings of various ECMs. It can be further exploited when coupled with investment financial analysis (see the earlier section in this chapter on such analysis) to ensure that the investment in first cost will pay for itself in annual energy savings. The whole design team should understand that the current state of energy modeling software is insufficient to predict actual energy use of a building but is adequate for comparing options to each other. It should be noted that there is no federal standard verifying the absolute accuracy of energy modeling engines as compared to real life in uncontrolled circumstances. Unlike many "repeatable" calculation techniques, error ranges are inconsistent based on systems used and user input, so there is no published percent accuracy associated with the algorithms. Additionally, the design team must acknowledge that the current software's algorithms are limited in many ways with regard to the manipulation of certain variables and control techniques and that energy modeling is as much an art as a science in terms of applying work-arounds for the limitations of the current code. All of this exposition is to say: "user beware." As a rule of thumb, results of at least a 5% relative energy savings arising from comparative energy models with ECMs applied to the same source file are probably a

Total Community Options Corporate Headquarters—A Case Study

Total Community Options's new 45,000 ft^2 corporate headquarters, located near the former Lowry Air Force Base in Denver, was completed in September 2010 and centralized several departments and business units that were spread across multiple locations throughout the city. The streamlined operational goals of the organization incited the design team to reduce the size of the building from the originally planned 60,000 ft^2 to the final 45,000 ft^2, with the construction budget for the eliminated 15,000 ft^2 put toward upgrading materials, systems, and amenities throughout the building and site.

Savings from long-term operating efficiencies of the building allow the nonprofit organization to spend more funds on programs and services. The design team was inspired to find solutions where a great deal of natural light and views were allowed. Operable windows were provided for employees to access natural ventilation, and the HVAC and lighting design were studied heavily to ensure that they were well honed and provided maximum occupant comfort. Perimeter spaces were designed with operable windows to give individuals control over the temperature in their area. Regularly occupied spaces on the interior without access to windows have adjustable air diffusers, which allow individuals to control their comfort by adjusting the amount of air supplied to a room.

Energy Conservation Measures

The Total Community Options office building achieves more than 41% energy and 43% energy cost savings through energy-efficiency designs as compared to an ASHRAE/IESNA Standard 90.1-2004 code-compliant building. With the 100 kW PV system, the building saves 70% in energy costs as compared to a typical office building. The building incorporates passive system design strategies, including optimized building orientation, increased insulation, daylighting, and natural ventilation, significantly cutting energy use before the mechanical system was evaluated. A variable-refrigerant-volume (VRV) HVAC system was incorporated that allows more control in all zones and also runs at variable speeds. In addition, a DOAS is controlled via CO_2 sensors in critical zones to reduce the conditioned outdoor air at occupancy rate. The DOAS unit has a variable-speed drive fan, direct evaporative cooling, and a modulating furnace.

Main Entry on South Face

Roof PV Panels

Architectural Features

- The site is oriented about 45 degrees from true north. A portion of the building is pulled out at an angle to face directly south.
- High-performance, low-e glazing was specified and windows were minimized as much as possible on the southwest façade. The WWR of 23% was optimized to provide the best possible daylighting while maintaining thermal performance.
- A three-story lobby and vestibule located at the main entrance serves as a buffer zone to the rest of the building and reduces the infiltration through the main entrance.
- Both rigid insulation and spray foam insulation were used in the wall system to achieve a true R-19 on the outside face of the metal stud exterior wall.
- The roof is covered with a white, reflective, single-ply roofing system that reflects light and heat away from the roof. This lowers the temperature of the roof and saves energy by requiring less cooling.

Daylight and Views

The office building was designed to maximize the impact of daylighting, both as an energy efficiency measure and as an occupant comfort strategy.
- The building floor plate is very narrow to allow daylight to reach almost all occupied areas within the building. Support spaces such as copy centers and mechanical rooms are placed farthest from glazing so regularly occupied spaces can take advantage of daylight and views.
- Glazing was added to interior rooms without perimeter windows to bring in additional natural light and provide views to the outdoors.
- High-performance, low-e glazing allows light in while preventing heat gain. Window coverings provide flexibility in controlling the amount of light. All perimeter spaces have shades or blinds to control glare and heat gain.
- In addition to view windows, dome-shaped tubular skylights were incorporated to concentrate sunlight and direct it deep into the floor plate, where daylight from view windows was minimal.

Rigid Insulation

Spray Foam Insulation

VRV HVAC System on Roof

(Continued on next page)

Interior Lighting

Lighting controls were extensively used. Perimeter spaces in the building have both daylight and occupancy sensors as well as stepped dimming lighting.

- When daylight levels are high, lighting fixtures near windows are automatically shut off using stepped dimming controls.
- Interior spaces have occupancy sensors to turn on lights when someone enters the room. Lighting fixtures automatically shut off after the room is unoccupied for a programmed period of time.
- LED task lighting provides individual controls so users can set light levels appropriate to their tasks.
- Open office areas use linear fluorescent direct/indirect fixtures, and the corridors have compact fluorescent down lights and decorative lighting.
- LED lighting is used for emergency exit signage, task lighting, and under-cabinet fixtures.

Renewable Energy

The 100 kW PV system provides 50% of the electricity for the building (a portion of the building also uses natural gas). There are two types of PV systems used on the building:

- The largest portions of PV panels are typical crystalline panels laid on the roof.
- Laminated glass PV panels are featured on the mansards on the front of the building to showcase the technology and create awareness among the building's occupants and visitors.

The Total Community Options office building is part of the Lowry Redevelopment—a former U.S. Air Force Base that has been redeveloped into a mixed-use, sustainable community. The Lowry Redevelopment has been recognized with several local and national smart growth awards, including the Governor's Smart Growth and Development Award and the 1999 Sustainable Community Award from the U.S. Conference of Mayors and National Association of Counties.

Low-E Glazing

Tubular Skylights

Case study provided by Chad Holtzinger, OZ Architecture of Denver

Photographs reprinted with permission of OZ Architecture of Denver

true indicator of measurable savings in real life. Anything less should be reviewed by the design team in a careful risk-management process.

The energy modeling process involves a very large amount of data input, and in some software programs it is extremely difficult to change geometry after the fact without reentering the entire model. As such, the design team should understand how it will choose to spend its limited energy modeling fee from the start and ensure that each model run is absolutely necessary to confirm beneficial direction.

There are many aspects of this Guide that prove energy-reduction benefits and are now best practice that do not need to be analyzed individually for cost-effectiveness. For instance, any nontechnical person would acknowledge that reduction in plug and lighting loads will reduce ultimate energy use given equivalent schedule of usage. In order to achieve 50% reduction in building energy use as compared to ASHRAE/IESNA Standard 90.1-2004 (ASHRAE 2004), energy modeling should be confirming the relative size of an already known benefit, not proving a bad position to resistant team members.

In light of the hours involved in developing a fully detailed ASHRAE/IESNA Standard 90.1-2004 whole-building energy modeling, decisions are often made without energy modeling input. Appendix D of this Guide includes an early-phase energy-use estimating method to help the team iterate the most basic of the considerations listed in these multidisciplinary integrated design tips to help point the team in the right direction before resorting to full energy modeling.

REFERENCES

ASHRAE. 2004. ANSI/ASHRAE/IESNA Standard 90.1-2004, *Energy Standard for Buildings Except Low-Rise Residential Buildings*. Atlanta: American Society of Heating, Refrigerating and Air-Conditioning.

ASHRAE. 2005. ASHRAE Guideline 0, *The Commissioning Process*. Atlanta: American Society of Heating, Refrigerating and Air-Conditioning.

ASHRAE. 2007a. ANSI/ASHRAE/IESNA Standard 90.1-2007, *Energy Standard for Buildings Except Low-Rise Residential Buildings*. Atlanta: American Society of Heating, Refrigerating and Air-Conditioning Engineers.

ASHRAE. 2007b. *ASHRAE Handbook—HVAC Applications*. Atlanta: American Society of Heating, Refrigerating and Air-Conditioning Engineers.

ASHRAE. 2007c. ASHRAE Guideline 1.1, *HVAC&R Technical Requirements for The Commissioning Process*. Atlanta: American Society of Heating, Refrigerating and Air-Conditioning.

ASHRAE. 2009a. ANSI/ASHRAE/USGBC/IES Standard 189.1-2009, *Standard for the Design of High-Performance Green Buildings Except Low-Rise Residential Buildings*. Atlanta: American Society of Heating, Refrigerating and Air-Conditioning.

ASHRAE. 2009b. *ASHRAE Handbook—Fundamentals*, I-P Edition. Atlanta: American Society of Heating, Refrigerating and Air-Conditioning.

ASHRAE. 2009c. *Indoor Air Quality Guide: Best Practices for Design, Construction, and Commissioning*. Atlanta: American Society of Heating, Refrigerating and Air-Conditioning Engineers.

ASHRAE. 2010a. ANSI/ASHRAE Standard 62.1-2010, *Ventilation for Acceptable Indoor Air Quality*. Atlanta: American Society of Heating, Refrigerating and Air-Conditioning.

ASHRAE. 2010b. ANSI/ASHRAE/IES Standard 90.1-2010, *Energy Standard for Buildings Except Low-Rise Residential Buildings*. Atlanta: American Society of Heating, Refrigerating and Air-Conditioning.

ASHRAE. 2010c. ANSI/ASHRAE Standard 55-2010, *Thermal Environmental Conditions for Human Occupancy*. Atlanta: American Society of Heating, Refrigerating and Air-Conditioning Engineers.

Briggs, R.S., R.G. Lucas, and Z.T. Taylor. 2002a. Climate classification for building energy codes and standards, PNNL Technical Paper final review draft. Richland, WA: Pacific Northwest National Laboratory.

Briggs, R.S., R.G. Lucas, and Z.T. Taylor. 2002b. Climate classification for building energy codes and standards: Part 1—Development process. *ASHRAE Transactions* 109(1).

Briggs, R.S., R.G. Lucas, and Z.T. Taylor. 2002c. Climate classification for building energy codes and standards: Part 2—Zone definitions, maps, and comparisons. *ASHRAE Transactions* 109(1).

CBSC. 2011. *California Energy Code*, Title 24, Part 6 of the California Code of Regulations. Sacramento: California Building Standards Commission.

CEC. 2008. California Commercial End-Use Survey (CEUS), www.energy.ca.gov/ceus. Sacramento: California Energy Commission.

CFR. 1992. National Appliance Energy Conservation Act. Code of Federal Regulations, Title 10, Chapter II, Part 430—Energy Efficiency Program for Certain Commercial and Industrial Equipment. Washington, DC: U.S. Government.

Colliver, D., R.S. Gates, T.F. Burks, and H. Zhang. 1997. Determination of the 0.4%, 1% and 2% annual occurrences of temperature and moisture and the 99% and 98% occurrences of temperature for 1400 national and international locations. Final Report, ASHRAE Research Project RP-890. Atlanta: American Society of Heating, Refrigerating and Air-Conditioning.

DOI. 1970. The National Atlas of the United States of America. Washington, DC: U.S. Department of Interior, Geological Survey.

EERE. 2010. Weather Data. EnergyPlus Energy Simulation Software. http://apps1.eere.energy.gov/buildings/energyplus/cfm/weather_data.cfm. Washington, DC: U.S. Department of Energy, Office of Energy Efficiency and Renewable Energy.

EIA. 2011. Commercial Buildings Energy Consumption Survey (CBECS), www.eia.doe.gov/emeu/cbecs. Washington, DC: U.S. Department of Energy, U.S. Energy Information Administration.

EPA. 2011. ENERGY STAR Portfolio Manager, www.energystar.gov/index.cfm?c=evaluate_performance.bus_portfoliomanager. Washington, DC: U.S. Environmental Protection Agency and U.S. Department of Energy.

Hart, R., S. Mangan, and W. Price. 2004. Who left the lights on? Typical load profiles in the 21st century. 2004 ACEEE Summer Study on Energy Efficiency in Buildings. Washington, DC: American Council for an Energy-Efficient Economy.

HMG. 2003. Windows and offices: A study of office worker performance and the indoor environment. California Energy Commission Technical Report P500-03-082-A-9. Gold River, CA: Heschong Mahone Group.

ICC. 2009. *International Energy Conservation Code*. Washington, DC: International Code Council.

IES. 2011. *The Lighting Handbook*, 10th ed. New York: Illuminating Engineering Society of North America.

Lobato, C., S. Pless, M. Sheppy, and P. Torcellini. 2011. Reducing plug and process loads for a large-scale, low-energy office building: NREL's Research Support Facility. *ASHRAE Transactions* 117(1):330–39.

Maniccia, D., and A. Tweed. 2000. *Occupancy Sensor Simulations and Energy Analysis for Commercial Buildings*. Troy, NY: Lighting Research Center, Renssaelaer Polytechnic Institute.

Marion, W., and S. Wilcox. 2008. *Users Manual for TMY3 Data Sets*, NREL/TP-581-43156. Golden, CO: National Renewable Energy Laboratory.

NFPA. 2011. NFPA 70, *National Electrical Code*. Quincy, MA: National Fire Protection Association.

NOAA. 2005. Climate maps of the United States. http://cdo.ncdc.noaa.gov/cgi-bin/climaps/climaps.pl. NOAA Satellite and Information Service. National Climatic Data Center. Washington, DC: National Oceanic and Atmospheric Administration.

NREL. 2005. Global Horizontal Solar Radiation - Annual. Image "solar_glo" available at www.nrel.gov/gis/images. Golden, CO: National Renewable Energy Laboratory.

Sanchez, M.C., C.A. Webber, R. Brown, J. Busch, M. Pinckard, and J. Roberson. 2007. Space heaters, computers, cell phone chargers: How plugged in are commercial buildings? LBNL-62397. *Proceedings of the 2006 ACEEE Summer Study on Energy Efficiency in Buildings*, August, Asilomar, CA.

Sethi, A., and T. Marseille. 2010. Old concepts, new tools: Case study—The Terry Thomas. *High Performing Buildings*, Summer:26–38.

Seibert, K.L. The right fit: Case study—CMTA office building. *High Performing Buildings*, Winter:48–59.

Thornton, B.A., W. Wang, M.D. Lane, M.I. Rosenberg, and B. Liu. 2009. *Technical Support Document: 50% Energy Savings Design Technology Packages for Medium Office Buildings*, PNNL-19004. Richland, WA: Pacific Northwest National Laboratory.

Thornton, B.A., W. Wang, Y. Huang, M.D. Lane, and B. Liu. 2010. *Technical Support Document: 50% Energy Savings for Small Office Buildings*, PNNL-19341. Richland, WA: Pacific Northwest National Laboratory.

USGBC. 2011. Leadership in Energy and Environmental Design (LEED) Green Building Rating System. Washington, DC: U.S. Green Building Council. www.usgbc.org/DisplayPage.aspx?CategoryID=19.

Design Strategies and Recommendations by Climate Zone

4

INTRODUCTION

Chapter 4 begins with general climate-related design strategies and follows with specific recommendations for each of the eight climate zones. The general design strategies need to be considered at the preliminary stages of building design. The chapter is segmented by climate conditions of temperature (hot, mild, and cold) and moisture (marine, humid, and dry). Each segment addresses the key issues associated with the climate, envelope, lighting, and heating, ventilating, and air conditioning (HVAC). The climate sections address conduction, solar loads, and moisture while the envelope sections address fenestration area, orientation, and shading. Daylighting is the focus of the lighting sections. The HVAC sections primarily address ventilation, economizers, and humidity control. No single design strategy applies universally to all of the climates. Each set of climate combinations needs to be analyzed separately.

The recommendations are presented in eight tables that contain the individual construction specifications per climate zone. Each table presents prescriptive recommendations for the following categories: building envelope, daylighting/lighting, plug loads, service water heating (SWH), and HVAC systems. Each category is subdivided into specific items that are then further subdivided into components. There is a prescriptive recommendation for each component along with a listing of relevant how-to tips. The how-to tips are presented in detail in Chapter 5. There are six HVAC options that provide the designer with a broad range of options, including packaged single-zone air-source heat pumps, water-source heat pumps with dedicated outdoor air systems (DOASs), radiant systems with DOASs, fan-coils with DOASs, variable air volume (VAV) direct expansion (DX), and VAV chilled water (CHW) (same as VAV DX but with some exceptions).

It is critical that the preliminary design follow the general design strategies since that is when the basic structure and form are set in terms of size, shape, and orientation. Once the basic design is set, the construction of office buildings needs to follow the prescriptive recommendations in order to achieve the 50% energy savings target.

CLIMATE-RELATED DESIGN STRATEGIES

The suggested approach is to first minimize the envelope heating and cooling loads. This has a double benefit: the energy use is reduced and a smaller-size HVAC system is needed to satisfy the reduced loads. Follow this by capitalizing on all of the daylighting opportunities.

HOT, HUMID CLIMATES (MIAMI, HOUSTON, ATLANTA)

Climate

The primary driving forces in hot, humid climates are conduction, solar loads through the fenestration, and significant cooling energy associated with removing indoor moisture due to people latent loads, ventilation, infiltration, and moisture ingress arising from storms or other disturbances coming off the Gulf of Mexico and Atlantic Ocean nearby.

Envelope

In these climates the fenestration area, orientation, and shading are paramount, as solar radiation intensities are among the highest in the continental U.S. The goal is to reduce the heat gain through the envelope as much as possible through strategic fenestration and shading placement. Generous verandas were historically applied to protect glazing, and modern designs often incorporate strategically placed external shading devices. Glazing type is usually double glazed in order to protect the low-e coating in the cavity to improve solar heat gain coefficient (SHGC) and to decouple the inner and outer faces of glass to reduce the risk of condensation on either side. SHGCs that are intentionally low are recommended—these can be achieved with interlayers or low-e coatings for spectrally selective transmission of sunlight to reduce the heat content while allowing light to enter. Care must be taken to reduce infiltration through the upper levels of the building envelope; positive building pressure control can help reduce infiltration and the related moisture. Cool roofs, which reduce solar heat absorption into the building, are also useful. It should be noted that these areas can experience high winds and hurricane storm impacts, which may direct the selection of wall, roof, and shading constructions as well as the selection of safety tempering of glazing products, all of which can affect the energy performance.

Lighting

Daylighting open office spaces on the north and south sides of the building works well, although the sizes and positions of windows should protect occupants from direct solar heat gain and direct glare. External shading devices will work at the southern façade; however, the choice to use these devices must be considered against the regular influx of storms and hurricanes through these areas. Internal light shelves with daylight glazing above (high visible transmittance [VT]) and view glazing below (low VT), along with horizontal blinds on the view glazing, can maximize daylighting potential and glare control without the need for external shading devices. As the sun does come north of the east-west line in early morning and late afternoon during the summer months, perpendicular fins may be necessary to reduce solar heat and glare even on northern façades.

HVAC

These climates experience average daily dew point temperatures higher than 50°F throughout much of the year. Because 72°F and 50% rh indoors has a dew point of about 52°F, it is clear that consistent humidity control is absolutely essential. A necessary strategy to maintain humidity control is proper dehumidification of all outdoor air (OA) for ventilation. This may be achieved with a 100% OAS with energy recovery wheels, and deep multirow cooling coils can provide sufficient dehumidification of ventilation air. It may also be achieved by mixed air (outdoor and recirculated) delivered with minimum flow setpoints and reheat (recovered if possible). Air-side and water-side economizers may have seasonal efficacy during times of lower dry-bulb and wet-bulb temperatures. During cool, humid conditions, DX systems must be carefully monitored and staged to prevent icing on the evaporator coils due to very low coil suction temperatures. Radiant cooling systems (ceilings or chilled beams) are possible in a tightly sealed envelope with excellent humidity control at the DOAS; however, many designers shy away from these systems because any substantial amount of infiltration can lead to internal condensation. Historically, natural ventilation schemes in these climates relied heavily on forced air movement to encourage rapid evaporation of sweat from exposed skin surfaces. All systems using heat rejection to the surrounding air should have increased heat rejection capabilities. During many months, these climates stay warm and humid overnight, so night cooling of interior thermal mass may not be effective. In hot, humid climates, indoor spaces, especially the upper floors, should be maintained at positive pressure with respect to the outdoor pressure to reduce infiltration.

HOT, DRY CLIMATES (PHOENIX, LOS ANGELES, LAS VEGAS)

Climate

The primary driving forces in hot, dry climates are conduction and solar loads through the fenestration and significant cooling energy associated with ventilation air.

Envelope

Strategic use of appropriately sized glazing, well-placed shading, increased insulation, and solar-reflective roofs and walls is highly recommended in order to reduce the influence of sun on the internal comfort and heat gain. Glazing type is usually double glazed in order to protect the low-e coating in the cavity to improve SHGC, and sometimes interpane blinds or prismatic elements are useful to offer shading while bouncing light in a certain direction. SHGCs that are intentionally low are recommended—these can be achieved with interlayers or low-e coatings for spectrally selective transmission of sunlight to reduce the heat content while allowing light to enter.

Lighting

Daylighting strategies that allow in light (particularly north light) without solar content are highly recommended, as these locations tend to have a high percentage of sunny days that might be exploited. The sizes and positions of windows should protect occupants from direct solar heat gain and glare, as the solar radiation in these areas is quite intense due to the relatively clear skies. External shading devices will work on the southern façade, and internal light shelves with daylight glazing above (high VT) and view glazing below (low VT), along with horizontal blinds on the view glazing, can maximize daylighting potential and glare control. In these southern climates, care must be taken even with north-facing glass, as the sun does come north of the east-west line in early morning and late afternoon during the summer months; perpendicular fins may be necessary to reduce solar heat and glare even on northern façades.

HVAC

These climates are often characterized by a large diurnal swing, so the use of free nighttime cooling should be explored in order to precool the interior surfaces or to perform nighttime cooling for thermal energy storage solutions (depending on the local demand charges). Heating is very minimal and can often be limited to the ventilation airstream with a small number of perimeter heating elements in zones with large expanses of glass. It should be noted that the air is very dry in these climates; however, it is not usual practice to provide active humidification. Many of these locations in the desert (Las Vegas, Phoenix) have seasonal monsoon cycles in late summer, and dehumidification capacity at the cooling coils should be sized for these extreme events. All of these locations can employ natural ventilation during the winter and shoulder seasons.

Air-side economizers are widely used. One system that is often used for energy savings is the indirect or indirect/direct evaporative cooler in the ventilation or economizer airstreams. Indirect evaporative cooling on a 100% OAS, coupled with radiant floors or ceilings, may be very effective in reducing energy use. In zones near the façade, however, usually a fan-coil unit or dedicated perimeter air-conditioning system is needed if the building is heavily glazed. If central plants are pursued, water-based cooling towers are very effective in these climates. For air-cooled DX equipment or air-cooled chillers, it is often an energy savings to consider evaporatively cooled condensers, especially when integrated to accept the exhaust from the indirect evaporative precoolers on the OA intakes. During the summer months, indoor spaces should be kept positive with respect to the outdoors in order to reduce infiltration.

MILD, HUMID CLIMATES (BALTIMORE)

Climate

The primary driving forces in mild, humid climates are conduction, solar loads through the fenestration, and significant cooling energy associated with removing indoor moisture due to ventilation, infiltration, and moisture ingress arising from summertime high humidities. During the winter, these areas are exposed to snow and freezing precipitation, so systems must be optimized to function efficiently in all seasons with sufficient responsiveness to ensure comfort during extreme swings in OA conditions.

Envelope

In these climates the goal is to reduce the heat gain and heat loss through the envelope's glazing as much as possible through strategic fenestration placement and sizing. Glazing type is usually double glazed in order to protect the low-e coating in the cavity to improve SHGC and to decouple the inner and outer faces of glass to reduce the risk of condensation on either side. SHGCs that are intentionally low are recommended if coupled with a low-e coating to improve U-factors during the winter season. Care must be taken with regard to minimizing infiltration and moisture ingress being driven through the building envelope. Cool roofs can be considered, but their usefulness will depend heavily on the sunniness of the local geography. It should be noted that these areas can experience high winds and Atlantic storm impacts, which may direct the selection of wall, roof, and shading constructions as well as the selection of safety tempering of glazing products, all of which can affect the energy performance.

Lighting

Daylighting strategies that allow in light (particularly north light) without solar content are highly recommended, as these locations tend to have a high percentage of sunny days that might be exploited. The sizes and positions of windows should protect occupants from direct solar heat gain and glare, as the solar radiation in these areas is quite intense due to the relatively clear skies. External shading devices will work at the southern façade, and light shelves at the east and west can bounce low-angle sun deep into the building footprint. Internal or external light shelves with daylight glazing above (high VT) and view glazing below (low VT), along with horizontal blinds on the view glazing, can maximize daylighting potential and glare control. In these southern climates, care must be taken even with north-facing glass, as the sun does come north of the east-west line in early morning and late afternoon during the summer months; perpendicular fins may be necessary to reduce solar heat and glare even on northern façades.

HVAC

These climates require dehumidification in the summer. Systems with energy recovery ventilation can precondition ventilation airflow and partially recover dehumidification or humidification energy that was used to condition the space. Air-side and water-side economizers can reduce mechanical cooling during the cold winter months. Most HVAC systems will work well in these climates as long as the building envelope and ventilation systems are designed to control moisture.

MILD, DRY CLIMATES (ALBUQUERQUE)

Climate

The primary driving forces in mild, dry climates are cooling and solar control. These climates are characterized by cool evenings and moderate diurnal swings, but significant heating is not required.

Envelope

Strategic use of appropriately sized glazing, well-placed shading, increased insulation, and solar-reflective roofs and walls is highly recommended in order to reduce the influence of sun on the internal comfort and heat gain. Glazing type is usually double glazed in order to protect the low-e coating in the cavity to improve SHGC. SHGCs that are intentionally low are recommended—these can be achieved with interlayers or low-e coatings for spectrally selective transmission of sunlight to reduce the heat content while allowing light to enter.

Lighting

Daylighting strategies that allow in light (particularly north light) without solar content are highly recommended, as these locations tend to have a high percentage of sunny days that might be exploited. The sizes and positions of windows should protect occupants from direct solar heat gain and glare, as the solar radiation in these areas is quite intense due to the relatively clear skies. External shading devices will work at the southern façade, and light shelves at the east and west can bounce low-angle sun deep into the building footprint. Internal or external light shelves with daylight glazing above (high VT) and view glazing below (low VT), along with horizontal blinds on the view glazing, can maximize daylighting potential and glare control. In these southern climates, care must be taken even with north-facing glass, as the sun does come north of the east-west line in early morning and late afternoon during the summer months; perpendicular fins may be necessary to reduce solar heat and glare even on northern façades.

HVAC

These climates are often characterized by a large diurnal swing, so the use of free nighttime cooling should be explored in order to precool the interior surfaces or to do nighttime cooling for thermal energy storage solutions (depending on the local demand charges). Heating is very minimal and can often be limited to the ventilation airstream with some perimeter heating in zones with large expanses of glass. Although the air is very dry in these climates, it is not usual practice to provide active humidification. These climates have extended periods of time when natural ventilation is effective.

Air-side economizers are widely used. Indirect or indirect/direct evaporative coolers are often used in the ventilation or economizer airstreams. Coupled with radiant floors or ceilings, they may be very effective in reducing energy use. In zones with heavily glazed façades, fan-coil units or dedicated perimeter air-conditioning systems are often needed. Evaporative heat-rejection systems, including cooling towers or evaporative condensers, are very effective in these climates. During the summer months, indoor spaces should be kept positive with respect to the outdoors in order to reduce infiltration.

MARINE CLIMATES (SAN FRANCISCO, SEATTLE)

Climate

The primary driving forces of marine climates are heating and solar control because outdoor conduction influences are minimal. These temperate climates are characterized by relatively stable temperature ranges in the cool-to-comfortable range, with early morning and evening moist conditions arising from lowered air temperatures given fairly stable and low dew points throughout the year created by mass flows from the cool nearby oceans.

Envelope

Adequate insulation for reduction of heating is necessary, including the use of double glazing in most locations.

Lighting

Daylighting is welcomed in these climates, as they tend to begin the morning rather gray with hopes of a burn-off. Larger expanses of glass are possible if they are double paned for heating control. Translucent exterior shading is often used in order to minimize the dark overhang of the solid shading and to maximize the amount of light entering the space. Internal or external light shelves with daylight glazing above (high VT) and view glazing below (low VT), along with horizontal blinds on the view glazing, can maximize daylighting potential and glare control.

HVAC

Effective system types for these climates exploit free cooling from moderate OA conditions and often provide minimal thermal intervention. These include mixed-mode solutions involving natural ventilation for certain portions of the year and radiant cooling/heating or passive chilled beams during the peak summer design days. It is usually not the practice to provide humidification in these climates.

Heating is almost always required for comfort, but is usually confined to perimeter heating elements with a small amount of ventilation-based boost heat. Usually, even during cold hours, office occupancies in these climates will exploit the air-side free-cooling economizer cycles at the main air handlers. Generally, water-side economizers are not effective in southern zones, but they can be beneficial in northern zones. Depending on ground-soil conductivity, ground-source heat pumps (GSHPs) have also been effective if total annual heating versus cooling is well-balanced.

COLD, DRY CLIMATES (DENVER, HELENA)

Climate

The primary driving forces in cold, dry climates are heat loss through the building envelope and attention to heating and cooling loads associated with ventilation air.

Envelope

Selection of insulation, care in the placement of vapor retarders, and insulating glazing are necessary in order to avoid internal drafts and condensation. Flat skylights are not recommended, as snow may build up in tall drifts. Vestibules at all entrances are virtually a necessity in order to avoid excessive waste of energy associated with infiltration. Infiltration and excessive building pressurization must be reduced as much as possible, as any moisture trapped in the wall constructions will experience a daily freeze-thaw cycle that can degrade wall integrity and effective R-values.

Lighting

Daylighting is welcomed in these climates. Usually there is a significant amount of exposure to clear skies during the longer summer days, and any outdoor light is welcomed in the winter. Larger expanses of glass are possible if they are double pane or even triple pane for heating control; however, special measures must be taken to reduce downdrafts and cold radiant surfaces at the windows. Internal or external light shelves with daylight glazing above (high

VT) and view glazing below (low VT), along with horizontal blinds on the view glazing, can maximize daylighting potential and glare control.

HVAC

These climates are heating dominated during the winter, with heating elements in any perimeter zone, freeze protection at all first-pass coils in ventilation air handlers, and humidification (with consideration of possible condensation on windows). Most buildings are held at just barely below neutral pressure throughout the colder months to prevent moist indoor air from being driven into the wall cavities, and in some instances where high humidity internal conditions are required, wall-drying systems must be installed. Building entries should incorporate heated, pressurized vestibules to prevent local infiltration of cold OA. These areas have very temperate shoulder seasons, so an air-side economizer cycle should be considered. In addition, indirect evaporative precoolers on the ventilation airstreams may be beneficial. Mixed-mode approaches with natural ventilation during some portions of the year and active heating and cooling during other portions can also be used with energy benefits. Although normally dry, some areas may experience sudden heavy rainfall for short periods, so activation of dehumidification capacity at the cooling coils requires speedy response times.

Depending on ground-soil conductivity, GSHPs have also been effective if total annual heating and cooling are well balanced. Evaporative heat rejection systems, including cooling towers or evaporative condensers, and water-side economizers are very effective in these climates.

COLD CLIMATES (CHICAGO, MINNEAPOLIS)

Climate

The primary driving forces in cold climates are heat loss through the building envelope, heat loss due to infiltration, and attention to heating and cooling loads associated with ventilation air. Because of the extreme cold, these loads tend to dwarf all other energy-use influences, especially during the winter months.

Envelope

Building envelope U-factors in cold climates are low to reduce conduction loads. It is also extremely important to reduce infiltration loads. When project teams attempt to positively pressurize the space, they must be careful to avoid too much exfiltration through leaky façades, as systems controlled to hold a fixed pressure differential will tend to increase the ventilation airflow to accommodate. Infiltration can be reduced by specifying and installing a high-quality continuous air barrier. Materials should be in compliance with or exceed the requirements of ASHRAE/IES Standard 90.1-2010 (ASHRAE 2010a). Insulation that expands into the wall cavity to provide the appropriate U-factor may also reduce infiltration and should be considered. In addition, envelope commissioning of the air barrier in cold climates may also be warranted. A vestibule is required by Standard 90.1-2010 in these climates and should be designed and constructed to avoid allowing the vestibule interior and exterior doors to be open simultaneously.

Lighting

Daylighting is welcomed in these climates. Usually there is a significant amount of exposure to clear skies during the longer summer days, and any outdoor light is welcomed in the winter. Larger expanses of glass are possible if they are double pane for heating control; however, special measures must be taken to reduce downdrafts and cold radiant surfaces at the windows. Internal or external light shelves with daylight glazing above (high VT) and view glazing below (low VT), along with horizontal blinds on the view glazing, can maximize daylighting potential and glare control.

HVAC

These climates are heating dominated during the winter, with heating elements in any perimeter zone, freeze protection at all first-pass coils in ventilation air handlers, and humidification (with consideration of possible condensation on windows). Building entries should incorporate heated, pressurized vestibules to prevent local infiltration of cold OA.

Ventilation loads can be significantly reduced by using a total energy recovery system. In cold climates, design teams must pay particular attention to the winter heat recovery attributes. While many small and medium office buildings do not incorporate humidification, latent energy recovery can mitigate extremely dry occupant spaces. High-efficiency energy recovery devices should be considered—especially in the most extreme climates.

To reduce heating equipment plant size yet maintain redundancy, select multiple smaller boilers to sum to the heating load, taking the energy recovery ventilator contributions into account. To ensure full energy recovery capacity at times with very cold temperatures, preheat the air entering the wheel on either the intake or the exhaust side.

In addition, appropriately reducing ventilation to only that required to satisfy ASHRAE Standard 62.1 requirements (ASHRAE 2010b) can be performed using demand-controlled ventilation (DCV) and ventilation reset at the central air handler or OA system. Ventilation reset is described by Murphy (2008), Stanke (2006, 2010), and Taylor (2006), among others.

A sometimes overlooked opportunity to reduce makeup air is to appropriately turn off or reduce airflow exhausted from spaces if allowed by local code.

While many low and medium-sized office buildings do not experience significant infiltration due to stack effect, vestibules or revolving doors can be considered to reduce infiltration in buildings of all heights.

During much of the year, air-side economizers can reduce or eliminate mechanical cooling if the system is sized to deliver enough airflow during these conditions. To reduce fan energy usage and ensure maximum economizer cooling capacity, bypass the energy recovery device while in economizer mode.

CLIMATE ZONE RECOMMENDATIONS

Users should determine the recommendations for their design and construction project by first locating the correct climate zone. The U.S. Department of Energy (DOE) has identified eight climate zones for the United States, with each defined by county borders, as shown in Figure 4-1. This Guide uses these DOE climate zones in defining energy recommendations that vary by climate. The definitions for the climate zones are provided in Appendix B so that the information can be applied outside the United States.

This chapter contains a unique set of energy-efficient recommendations for each climate zone. The recommendation tables represent *a way*, but *not the only way*, for reaching the 50% energy savings target over ASHRAE/IESNA Standard 90.1–2004 (ASHRAE 2004). Other approaches may also save energy, and Chapter 3 is intended to summarize the strategic multi-disciplinary decisions that may generate other viable integrated options; confirmation of energy savings for those uniquely designed systems is left to the design team. The user should note that the recommendation tables do not include all of the components listed in Standard 90.1 since the Guide focuses only on the primary energy systems within a building.

BONUS SAVINGS

Chapter 5 provides additional recommendations and strategies for savings for toplighting, natural ventilation, and renewable energy that are over and above the 50% savings recommendations contained in the eight climate regions.

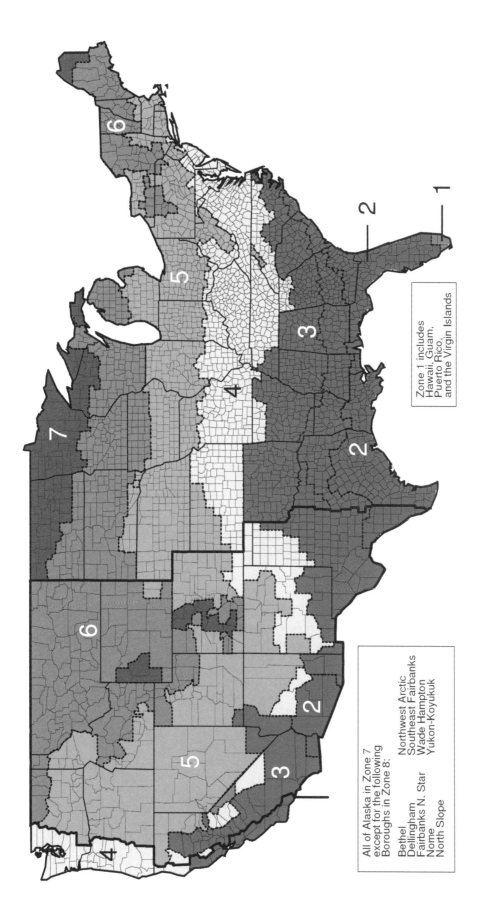

Figure 4-1 U.S. Map Showing the DOE Climate Zones (Briggs et al. 2003)

When a recommendation is provided, the recommended value differs from the requirements in Standard 90.1–2004. When "No recommendation" is indicated, the user must meet the more stringent of either the applicable version of Standard 90.1 or the local code requirements.

Each of the recommendation tables includes a set of common items arranged by building subsystem: envelope, daylighting/lighting, plug loads, SWH, and HVAC. Recommendations are included for each item, or subsystem, by component within that subsystem. For some subsystems, recommendations depend on the construction type. For example, insulation values are given for mass and steel-framed and wood-framed wall types. For other subsystems, recommendations are given for each subsystem attribute. For example, vertical fenestration recommendations are given for thermal transmittance, SHGC, and exterior sun control.

The "How-To Tips" column in each table lists references to how-to tips for implementing the recommended criteria. The tips are found in Chapter 5 under separate sections coded for envelope (EN), daylighting (DL), electric lighting (EL), plug loads (PL), service water heating systems and equipment (WH), HVAC systems and equipment (HV), and quality assurance (QA). In addition to how-to tips that represent good practice for design and maintenance suggestions, these tips include cautions for what to avoid. Important QA considerations and recommendations are also given for the building design, construction, and post-occupancy phases. Note that each tip is tied to the applicable climate zone in Chapter 4. The final column is provided as a simple checklist to identify the recommendations being used for a specific building design and construction.

The recommendations presented are minimum, maximum, or specific values (which are both the minimum and maximum values).

Minimum values include values for the following:

- R-value
- Solar Reflectance Index (SRI)
- Visible transmittance (VT)
- Vertical fenestration effective aperture (EA)
- Interior surface average reflectance
- Mean lumens per watt (LPW)
- Gas water heater or boiler efficiency
- Thermal efficiency (E_t)
- Energy factor (EF)
- Energy efficiency ratio (EER)
- Integrated energy efficiency ratio (IEER)
- Integrated part-load value (IPLV)
- Coefficient of performance (COP)
- Energy recovery effectiveness
- Fan or motor efficiency
- Duct or pipe insulation thickness

Maximum values include values for the following:

- Fenestration and door U-factors
- Fenestration solar heat gain coefficient (SHGC)
- Lighting power density (LPD)
- Fan input power per cubic foot per minute of supply airflow
- Window-to-wall ratio (WWR)
- External static pressure (ESP)
- Duct friction rate

Zone 1

Florida
  Broward
  Miami-Dade
  Monroe

Guam

Hawaii

Puerto Rico

U.S. Virgin Islands

Climate Zone 1 Recommendation Table for Small to Medium Office Buildings

	Item	Component	Recommendation	How-to Tips	✓
Envelope	Roofs	Insulation entirely above deck	R-20.0 c.i.	EN2, 17, 19, 21–22	
		Attic and other	R-38.0	EN3, 17, 19, 20–21	
		Metal building	R-19.0 + R-10.0 FC	EN4, 17, 19, 21	
		SRI	78	EN1	
	Walls	Mass (HC > 7 Btu/ft^2)	R-5.7 c.i.	EN5, 17, 19, 21	
		Steel framed	R-13.0 + R-7.5 c.i.	EN6, 17, 19, 21	
		Wood framed and other	R-13.0	EN7, 17, 19, 21	
		Metal building	R-0.0 + R-9.8 c.i.	EN8, 17, 19, 21	
		Below-grade walls	No recommendation	EN9, 17, 19, 21–22	
	Floors	Mass	R-4.2 c.i.	EN10, 17, 19, 21	
		Steel joint	R-19.0	EN11, 17, 19, 21	
		Wood framed and other	R-19.0	EN11, 17, 19, 21	
	Slabs	Unheated	No recommendation	None	
		Heated	R-7.5 for 12 in.	EN13–14, 17, 19, 21–22	
	Doors	Swinging	U-0.70	EN15, 17–18	
		Nonswinging	U-1.45	EN16–17	
	Vestibules	At building entrance	No recommendation	None	
	Continuous Air Barriers	Continuous air barrier	Entire building envelope	EN17	
	Vertical Fenestration	WWR	20% to 40%	EN25; DL6	
		Window orientation	Area of W and E windows each less than area of S windows (N in southern hemisphere)	EN28, 30	
		Exterior sun control (S, E, and W only)	PF-0.5	EN26	
		Thermal transmittance	Nonmetal framing windows = U-0.56 Metal framing windows = U-0.65	EN23–24	
		SHGC	Nonmetal framing windows = 0.25 Metal framing windows = 0.25	EN 24, 29; DL12	
Daylighting / Lighting	Daylighting	Light-to-solar-gain ratio	Minimum VT/SHGC = 1.10	EN34–42	
		Vertical fenestration EA	0.08	DL7–11	
	Interior Finishes	Interior surface average reflectance	Ceilings = 80% Wall surfaces = 70% Open office partitions = 50%	DL14–15; EL4	
		Open office partitions parallel to window walls	Total partition height = 36 in. maximum - or - Partition above desk height = min 50% translucent	DL4, 10; EL2, 4	
	Interior Lighting	LPD	0.75 W/ft^2	DL1–5; EL2–3, 5, 12–19	
		24-hour lighting LPD	0.075 W/ft^2	EL20	
		Light source lamp efficacy (mean LPW)	T8 and T5 lamps > 2 ft = 92 T8 and T5 lamps ≤ 2 ft = 85 All other > 50	EL6–8	
		Ballasts	4 ft T8 lamp nondimming applications = NEMA Premium instant start 4 ft T8 lamp dimming applications = NEMA Premium program start Fluorescent and HID sources = electronic	EL8	
		Controls for daylight harvesting in open offices—locate on N and S sides of bldg	Dim all general fluorescent lights within primary and secondary daylight zones of open offices	DL16–20; EL1, 11	
		Automatic controls	Auto ON to 50% = private offices, conference and meeting rooms, lounge and break rooms, copy rooms, storage rooms Auto ON occupancy sensors = restrooms, electrical/ mechanical rooms, open and private office task lighting Time switch control = all other spaces	DL13; EL1, 5, 9–10	
	Exterior Lighting	Façade and Landscape lighting	LPD = 0.075 W/ft^2 in LZ3 and LZ4, 0.05 W/ft^2 in LZ2 Controls = auto OFF between 12 am and 6 am	EL23	
		Parking lots and drives	LPD = 0.1 W/ft^2 in LZ3 and LZ4, 0.06 W/ft^2 in LZ2 Controls = auto reduce to 25% (12 am to 6 am)	EL21, 24–25	
		Walkways, plazas, and special feature areas	LPD = 0.16 W/ft^2 LZ3 and LZ4, 0.14 W/ft^2 in LZ2 Controls = auto reduce to 25% (12 am to 6 am)	EL22, 24–25	
		All other exterior lighting	LPD = follow Standard 90.1-2010 Controls = auto reduce to 25% (12 am to 6 am)	EL24–25	
Plug Loads	Equipment Choices	Laptop computers	Minimum 2/3 of total computers	PL1–2	
		ENERGY STAR equipment	For all computers, equipment, and appliances	PL1–2	
		Equipment power density	For all computers, equipment, and appliances	PL1–2, 4–6	
	Controls	Computer power control	Network control with power saving modes and control OFF during unoccupied hours	PL3	
		Occupancy sensors	Desk plug strip occupancy sensors	PL3	
		Timer switches	Water coolers and coffee makers control OFF during unoccupied hours	PL3	
		Vending machine control	Yes	PL3	

*Note: Where the table says "No recommendation," the user must meet the more stringent of either the applicable version of ASHRAE/IES Standard 90.1 or the local code requirements.

Climate Zone 1 Recommendation Table for Small to Medium Office Buildings *(Continued)*

	Item	Component	Recommendation	How-to Tips	✓
SWH	SWH	Gas water heater efficiency	Condensing water heaters = 90% efficiency	WH1–5	
		Electric storage EF (≤12 kW, ≥20 gal)	EF > 0.99 – 0.0012 × volume (see Table 5-7)	WH1–5	
		Point-of-use heater selection	0.81 EF or 81% E_t	WH1–5	
		Electric heat pump water heater efficiency	COP 3.0 (interior heat source)	WH1–5	
		Pipe insulation (d < 1½ in. / d ≥ 1½ in.)	1 in. / 1 ½ in.	WH6	
HVAC	Packaged Single-Zone Air-Source Heat Pumps	Cooling and heating efficiency	See Table 5-8 for efficiency	HV3	
		ESP	0.7 in. w.c.	HV3	
		DOAS air-source heat pump	Yes	HV10	
		DOAS heating and cooling efficiency	See Table 5-10 for efficiency	HV10–11	
		DOAS energy recovery	Yes, see Table 5-11 for effectiveness	HV12	
		DOAS fan—ESP	1.5 in. w.c. maximum	HV20	
		Unit size	5 tons or less	None	
		Single-stage cooling and heating efficiency	Cooling = 16.4 EER Heating = 5.2 COP	HV4	
		Two-stage cooling and heating efficiency	Cooling part load = 17.6 EER Cooling full load = 15.0 EER Heating part load = 5.7 COP Heating full load = 5.0 COP	HV4	
	WSHPs with DOAS	WSHP fan—ESP	0.5 in. w.c.	HV4, 20	
		Condensing boiler efficiency	90%	HV14, 30	
		DOAS water-to-water heat pump	See Table 5-10 for efficiency	HV10–11	
		DOAS variable airflow with DCV	Yes	HV17–18	
		DOAS energy recovery	Yes, see Table 5-11 for effectiveness	HV12	
		DOAS fan and motor	65% mechanical efficiency Motor efficiency per Standard 90.1-2010, Table 10.8B VSD efficiency = 95%	HV23	
	VAV DX with Indirect Gas-Fired or Elextric Internal Heat and Electric Perimeter Heat	DX efficiency	See Table 5-9 for efficiency	HV6	
		Low-temperature air supply and SAT reset	50°F to 58°F	HV6, 11, 25, 31	
		Perimeter convector heat source	Electric	HV6	
		Gas furnace in DX units	No recommendation	HV6	
		Economizer	No recommendation	None	
		Energy recovery	Yes, see Table 5-11 for effectiveness	HV12	
		Indirect evaporative cooling	No recommendation	None	
		Demand control and ventilation reset	Yes	HV17–18	
		ESP	2.0 in. w.c.	HV7, 20	
	VAV CHW (same as VAV DX except...)	Air-cooled chiller efficiency	10 EER	HV14, 29	
		Air-cooled chiller IPLV	12.5 IPLV < 150 tons, 12.75 IPLV ≥ 150 tons	HV14	
		Variable-speed pumping	Yes	HV29	
		Maximum fan power	0.72 W/cfm	HV7, 23, 31	
	Fan-Coils with DOAS	Air-cooled chiller efficiency	10 EER	HV14, 29	
		Air-cooled chiller IPLV	12.5 IPLV < 150 tons 12.75 IPLV ≥ 150 tons	HV14	
		Condensing boiler efficiency	90%	HV14, 30	
		Variable-speed pumping	Yes	HV29–30	
		VAV fan-coil units	Yes	HV8	
		Fan-coil unit fan power	0.30 W/cfm	HV8	
		DOAS chilled-water and hot-water coils served by same plant as fan-coils	Yes	HV10	
		DOAS variable airflow with DCV	Yes	HV10–11, 17	
		DOAS energy recovery	Yes, see Table 5-11 for effectiveness	HV12	
		DOAS fan and motor	65% mechanical efficiency, motor efficiency Standard 90.1-2010, Table 10.8B	HV23	
	Radiant Systems with DOAS	Air-cooled chiller full-load efficiency	10 EER	HV14, 29	
		Air-cooled chiller IPLV	12.5 IPLV < 150 tons, 12.75 IPLV ≥ 150 tons	HV14	
		Condensing boiler efficiency	90%	HV14, 30	
		DOAS heating and cooling efficiency	See Table 5-10 for efficiency	HV10–11	
		DOAS variable airflow with DCV	Yes	HV10–11, 17–18	
		DOAS energy recovery	Yes, see Table 5-11 for effectiveness	HV12	
		DOAS fan—ESP	1.5 in. w.c.	HV20	
	Ducts and Dampers	OA damper	Motorized damper	HV16	
		Friction rate	0.08 in./100 ft	HV20	
		Sealing	Seal Class B	HV22	
		Location	Interior only	HV20	
		Insulation level	R-6.0	HV21	

*Note: Where the table says "No recommendation," the user must meet the more stringent of either the applicable version of ASHRAE/IES Standard 90.1 or the local code requirements.

Zone 2

Alabama

Baldwin
Mobile

Arizona

La Paz
Maricopa
Pima
Pinal
Yuma

California

Imperial

Florida

Alachua
Baker
Bay
Bradford
Brevard
Calhoun
Charlotte
Citrus
Clay
Collier
Columbia
DeSoto
Dixie
Duval
Escambia
Flagler
Franklin
Gadsden
Gilchrist
Glades
Gulf

Hamilton
Hardee
Hendry
Hernando
Highlands
Hillsborough
Holmes
Indian River
Jackson
Jefferson
Lafayette
Lake
Lee
Leon
Levy
Liberty
Madison
Manatee
Marion
Martin
Nassau
Okaloosa
Okeechobee
Orange
Osceola
Palm Beach
Pasco
Pinellas
Polk
Putnam
Santa Rosa
Sarasota
Seminole
St. Johns
St. Lucie
Sumter
Suwannee
Taylor

Union
Volusia
Wakulla
Walton
Washington

Georgia

Appling
Atkinson
Bacon
Baker
Berrien
Brantley
Brooks
Bryan
Camden
Charlton
Chatham
Clinch
Colquitt
Cook
Decatur
Echols
Effingham
Evans
Glynn
Grady
Jeff Davis
Lanier
Liberty
Long
Lowndes
McIntosh
Miller
Mitchell
Pierce
Seminole
Tattnall

Thomas
Toombs
Ware
Wayne

Louisiana

Acadia
Allen
Ascension
Assumption
Avoyelles
Beauregard
Calcasieu
Cameron
East Baton
Rouge
East Feliciana
Evangeline
Iberia
Iberville
Jefferson
Jefferson Davis
Lafayette
Lafourche
Livingston
Orleans
Plaquemines
Pointe Coupee
Rapides
St. Bernard
St. Charles
St. Helena
St. James
St. John the
Baptist
St. Landry
St. Martin
St. Mary

St. Tammany
Tangipahoa
Terrebonne
Vermilion
Washington
West Baton
Rouge
West Feliciana

Mississippi

Hancock
Harrison
Jackson
Pearl River
Stone

Texas

Anderson
Angelina
Aransas
Atascosa
Austin
Bandera
Bastrop
Bee
Bell
Bexar
Bosque
Brazoria
Brazos
Brooks
Burleson
Caldwell
Calhoun
Cameron
Chambers
Cherokee

Colorado
Comal
Coryell
DeWitt
Dimmit
Duval
Edwards
Falls
Fayette
Fort Bend
Freestone
Frio
Galveston
Goliad
Gonzales
Grimes
Guadalupe
Hardin
Harris
Hays
Hidalgo
Hill
Houston
Jackson
Jasper
Jefferson
Jim Hogg
Jim Wells
Karnes
Kenedy
Kinney
Kleberg
La Salle
Lavaca
Lee
Leon
Liberty
Limestone

Live Oak
Madison
Matagorda
Maverick
McLennan
McMullen
Medina
Milam
Montgomery
Newton
Nueces
Orange
Polk
Real
Refugio
Robertson
San Jacinto
San Patricio
Starr
Travis
Trinity
Tyler
Uvalde
Val Verde
Victoria
Walker
Waller
Washington
Webb
Wharton
Willacy
Williamson
Wilson
Zapata
Zavala

Climate Zone 2 Recommendation Table for Small to Medium Office Buildings

	Item	Component	Recommendation	How-To Tips	✓
Envelope	Roofs	Insulation entirely above deck	R-25.0 c.i.	EN2, 17, 19, 21–22	
		Attic and other	R-38.0	EN3, 17, 19, 20–21	
		Metal building	R-19.0 + R-10.0 FC	EN4, 17, 19, 21	
		SRI	78	EN1	
	Walls	Mass (HC > 7 Btu/ft^2)	R-7.6 c.i.	EN5, 17, 19, 21	
		Steel framed	R-13.0 + R-7.5 c.i.	EN6, 17, 19, 21	
		Wood framed and other	R-13.0 + R-3.8 c.i.	EN7, 17, 19, 21	
		Metal building	R-0.0 + R-9.8 c.i.	EN8, 17, 19, 21	
		Below-grade walls	No recommendation	EN9, 17, 19, 21–22	
	Floors	Mass	R-10.4 c.i.	EN10, 17, 19, 21	
		Steel joint	R-30.0	EN11, 17, 19, 21	
		Wood framed and other	R-30.0	EN11, 17, 19, 21	
	Slabs	Unheated	No recommendation	None	
		Heated	R-10.0 for 24 in.	EN13–14, 17, 19, 21–22	
	Doors	Swinging	U-0.70	EN15, 17–18	
		Nonswinging	U-0.50	EN16–17	
	Vestibules	At building entrance	No recommendation	None	
	Continuous Air Barriers	Continuous air barrier	Entire building envelope	EN17	
	Vertical Fenestration	WWR	20% to 40%	EN25; DL6	
		Window orientation	Area of W and E windows each less than area of S windows (N in southern hemisphere)	EN28, 30	
		Exterior sun control (S, E, and W only)	PF-0.5	EN26	
		Thermal transmittance	Nonmetal framing windows = U-0.45 Metal framing windows = U-0.65	EN23–24	
		SHGC	Nonmetal framing windows = 0.25 Metal framing windows = 0.25	EN 24, 29; DL12	
Daylighting / Lighting	Daylighting	Light-to-solar-gain ratio	Minimum VT/SHGC = 1.10	EN34–42	
		Vertical fenestration EA	0.08	DL7–11	
	Interior Finishes	Interior surface average reflectance	Ceilings = 80% Wall surfaces = 70% Open office partitions = 50%	DL14–15; EL4	
		Open office partitions parallel to window walls	Total partition height = 36 in. maximum - or - Partition above desk height = min 50% translucent	DL4, 10; EL2, 4	
	Interior Lighting	LPD	0.75 W/ft^2	DL1–5; EL2–3, 5, 12–19	
		24-hour lighting LPD	0.075 W/ft^2	EL20	
		Light source lamp efficacy (mean LPW)	T8 and T5 lamps > 2 ft = 92 T8 and T5 lamps ≤ 2 ft = 85 All other > 50	EL6–8	
		Ballasts	4 ft T8 lamp nondimming applications = NEMA Premium instant start 4 ft T8 lamp dimming applications = NEMA Premium program start Fluorescent and HID sources = electronic	EL8	
		Controls for daylight harvesting in open offices—locate on N and S sides of bldg	Dim all general fluorescent lights within primary and secondary daylight zones of open offices	DL16–20; EL1, 11	
		Automatic controls	Auto ON to 50% = private offices, conference and meeting rooms, lounge and break rooms, copy rooms, storage rooms Auto ON occupancy sensors = restrooms, electrical/mechanical rooms, open and private office task lighting Time switch control = all other spaces	DL13; EL1, 5, 9–10	
	Exterior Lighting	Façade and landscape lighting	LPD = 0.075 W/ft^2 in LZ3 and LZ4, 0.05 W/ft^2 in LZ2 Controls = auto OFF between 12 am and 6 am	EL23	
		Parking lots and drives	LPD = 0.1 W/ft^2 in LZ3 and LZ4, 0.06 W/ft^2 in LZ2 Controls = auto reduce to 25% (12 am to 6 am)	EL21, 24–25	
		Walkways, plazas, and special feature areas	LPD = 0.16 W/ft^2 LZ3 and LZ4, 0.14 W/ft^2 in LZ2 Controls = auto reduce to 25% (12 am to 6 am)	EL22, 24–25	
		All other exterior lighting	LPD = follow Standard 90.1-2010 Controls = auto reduce to 25% (12 am to 6 am)	EL24–25	
Plug Loads	Equipment Choices	Laptop computers	Minimum 2/3 of total computers	PL1–2	
		ENERGY STAR equipment	For all computers, equipment, and appliances	PL1–2	
		Equipment power density	For all computers, equipment, and appliances	PL1–2, 4–6	
	Controls	Computer power control	Network control with power saving modes and control OFF during unoccupied hours	PL3	
		Occupancy sensors	Desk plug strip occupancy sensors	PL3	
		Timer switches	Water coolers and coffee makers control OFF during unoccupied hours	PL3	
		Vending machine control	Yes	PL3	

*Note: Where the table says "No recommendation," the user must meet the more stringent of either the applicable version of ASHRAE/IES Standard 90.1 or the local code requirements.

Climate Zone 2 Recommendation Table for Small to Medium Office Buildings *(Continued)*

Item	Component	Recommendation	How-To Tips	✓
SWH — SWH	Gas water heater efficiency	Condensing water heaters = 90% efficiency	WH1–5	
	Electric storage EF (≤12 kW, ≥20 gal)	EF > 0.99 – 0.0012 × volume (see Table 5-7)	WH1–5	
	Point-of-use heater selection	0.81 EF or 81% E_t	WH1–5	
	Electric heat pump water heater efficiency	COP 3.0 (interior heat source)	WH1–5	
	Pipe insulation (d < 1½ in. / d ≥ 1½ in.)	1 in. / 1 ½ in.	WH6	
HVAC — Packaged Single-Zone Air-Source Heat Pumps	Cooling and heating efficiency	See Table 5-8 for efficiency	HV3	
	ESP	0.7 in. w.c.	HV3	
	DOAS air-source heat pump	Yes	HV10	
	DOAS heating and cooling efficiency	See Table 5-10 for efficiency	HV10–11	
	DOAS energy recovery	Yes, see Table 5-11 for effectiveness	HV12	
	DOAS fan—ESP	1.5 in. w.c. maximum	HV20	
WSHPs with DOAS	Unit size	5 tons or less	None	
	Single-stage cooling and heating efficiency	Cooling = 16.4 EER Heating = 5.2 COP	HV4	
	Two-stage cooling and heating efficiency	Cooling part load = 17.6 EER Cooling full load = 15.0 EER Heating part load = 5.7 COP Heating full load = 5.0 COP	HV4	
	WSHP fan—ESP	0.5 in. w.c.	HV4, 20	
	Condensing boiler efficiency	90%	HV14, 30	
	DOAS water-to-water heat pump	See Table 5-10 for efficiency	HV10–11	
	DOAS variable airflow with DCV	Yes	HV17–18	
	DOAS energy recovery	Yes, see Table 5-11 for effectiveness	HV12	
	DOAS fan and motor	65% mechanical efficiency Motor efficiency per Standard 90.1-2010, Table 10.8B VSD efficiency = 95%	HV23	
VAV DX with Indirect Gas-Fired or Elextric Internal Heat and Electric Perimeter Heat	DX efficiency	See Table 5-9 for efficiency	HV6	
	Low-temperature air supply and SAT reset	50°F to 58°F	HV6, 11, 25, 31	
	Perimeter convector heat source	Electric	HV6	
	Gas furnace in DX units	No recommendation	HV6	
	Economizer	≥54,000 Btu/h, differential enthalpy control (Climate zone 2A) ≥54,000 Btu/h, differential dry-bulb control (Climate zone 2B)	HV6, 16	
	Energy recovery	Yes, see Table 5-11 for effectiveness	HV12	
	Indirect evaporative cooling	Climate zone 2B only	HV13, 36	
	Demand control and ventilation reset	Yes	HV17–18	
	ESP	2.0 in. w.c.	HV7, 20	
VAV CHW (same as VAV DX except...)	Air-cooled chiller efficiency	10 EER	HV14, 29	
	Air-cooled chiller IPLV	12.5 IPLV < 150 tons, 12.75 IPLV ≥ 150 tons	HV14	
	Variable-speed pumping	Yes	HV29	
	Maximum fan power	0.72 W/cfm	HV7, 23, 31	
Fan-Coils with DOAS	Air-cooled chiller efficiency	10 EER	HV14, 29	
	Air-cooled chiller IPLV	12.5 IPLV < 150 tons 12.75 IPLV ≥ 150 tons	HV14	
	Condensing boiler efficiency	90%	HV14, 30	
	Variable-speed pumping	Yes	HV29–30	
	VAV fan-coil units	Yes	HV8	
	Fan-coil unit fan power	0.30 W/cfm	HV8	
	DOAS chilled-water and hot-water coils served by same plant as fan-coils	Yes	HV10	
	DOAS variable airflow with DCV	Yes	HV10–11, 17	
	DOAS energy recovery	Yes, see Table 5-11 for effectiveness	HV12	
	DOAS fan and motor	65% mechanical efficiency, motor efficiency Standard 90.1-2010, Table 10.8B	HV23	
Radiant Systems with DOAS	Air-cooled chiller full-load efficiency	10 EER	HV14, 29	
	Air-cooled chiller IPLV	12.5 IPLV < 150 tons, 12.75 IPLV ≥ 150 tons	HV14	
	Condensing boiler efficiency	90%	HV14, 30	
	DOAS heating and cooling efficiency	See Table 5-10 for efficiency	HV10–11	
	DOAS variable airflow with DCV	Yes	HV10–11, 17–18	
	DOAS energy recovery	Yes, see Table 5-11 for effectiveness	HV12	
	DOAS fan—ESP	1.5 in. w.c.	HV20	
Ducts and Dampers	OA damper	Motorized damper	HV16	
	Friction rate	0.08 in./100 ft	HV20	
	Sealing	Seal Class B	HV22	
	Location	Interior only	HV20	
	Insulation level	R-6.0	HV21	

*Note: Where the table says "No recommendation," the user must meet the more stringent of either the applicable version of ASHRAE/IES Standard 90.1 or the local code requirements.

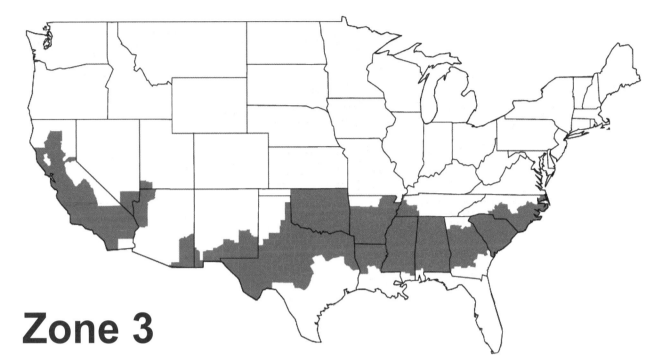

Zone 3

Alabama

All counties except:
Baldwin
Mobile

Arizona

Cochise
Graham
Greenlee
Mohave
Santa Cruz

Arkansas

All counties except:
Baxter
Benton
Boone
Carroll
Fulton
Izard
Madison
Morion
Newton
Searcy
Stone
Washington

California

All counties except:
Alpine
Amador
Calaveras
Del Norte
El Dorado
Humboldt
Imperial
Inyo
Lake
Lassen
Mariposa
Modoc
Mono
Nevada
Plumas
Sierra
Siskiyou
Trinity
Tuolumne

Georgia

All counties except:
Appling
Atkinson
Bacon
Baker
Banks
Berrien
Brantley
Brooks
Bryan
Catoosa
Camden
Charlton
Chatham
Chattooga
Clinch
Colquitt
Cook
Dade
Dawson
Decatur
Echols
Effingham
Evans
Fannin
Floyd
Franklin
Gilmer
Glynn
Gordon
Grady
Habersham
Hall
Jeff Davis
Lanier
Liberty
Long
Lowndes
Lumpkin
McIntosh
Miller
Mitchell
Murray
Pickens
Pierce
Rabun
Seminole
Stephens
Tattnall
Thomas
Toombs
Towns
Union

Walker
Ware
Wayne
White
Whitfield

Louisiana

Bienville
Bossier
Caddo
Caldwell
Catahoula
Claiborne
Concordia
De Soto
East Carroll
Franklin
Grant
Jackson
La Salle
Lincoln
Madison
Morehouse
Natchitoches
Ouachita
Red River
Richland
Sabine
Tensas
Union
Vernon
Webster
West Carroll
Winn

Mississippi

All counties except:
Hancock
Harrison
Jackson
Pearl River
Stone

New Mexico

Chaves
Dona Ana
Eddy
Hidalgo
Lea
Luna
Otero

Nevada

Clark

Texas

Andrews
Archer
Baylor
Blanco
Borden
Bowie
Brewster
Brown
Burnet
Callahan
Camp
Cass
Childress
Clay
Coke
Coleman
Collingsworth
Collin
Comanche
Concho
Cottle
Cooke
Crane
Crockett
Crosby
Culberson
Dallas
Dawson
Delta
Denton
Dickens
Eastland
Ector
El Paso
Ellis
Erath
Fannin
Fisher
Foard
Franklin
Gaines
Garza
Gillespie
Glasscock
Grayson
Gregg
Hall
Hamilton
Hardeman

Harrison
Haskell
Hemphill
Henderson
Hood
Hopkins
Howard
Hudspeth
Hunt
Irion
Jack
Jeff Davis
Johnson
Jones
Kaufman
Kendall
Kent
Kerr
Kimble
King
Knox
Lamar
Lampasas
Llano
Loving
Lubbock
Lynn
Marion
Martin
Mason
McCulloch
Menard
Midland
Mills
Mitchell
Montague
Morris
Motley
Nacogdoches
Navarro
Nolan
Palo Pinto
Panola
Parker
Pecos
Presidio
Rains
Reagan
Reeves
Red River
Rockwall
Runnels
Rusk
Sabine
San Augustine

San Saba
Schleicher
Scurry
Shackelford
Shelby
Smith
Somervell
Stephens
Sterling
Stonewall
Sutton
Tarrant
Taylor
Terrell
Terry
Throckmorton
Titus
Tom Green
Upshur
Upton
Van Zandt
Ward
Wheeler
Wichita
Wilbarger
Winkler
Wise
Wood
Young

Utah

Washington

North Carolina

Anson
Beaufort
Bladen
Brunswick
Cabarrus
Camden
Carteret
Chowan
Columbus
Craven
Cumberland
Currituck
Dare
Davidson
Duplin
Edgecombe
Gaston
Greene
Hoke
Hyde

Johnston
Jones
Lenoir
Martin
Mecklenburg
Montgomery
Moore
New Hanover
Onslow
Pamlico
Pasquotank
Pender
Perquimans
Pitt
Randolph
Richmond
Robeson
Rowan
Sampson
Scotland
Stanly
Tyrrell
Union
Washington
Wayne
Wilson

Oklahoma

All counties except:
Beaver
Cimarron
Texas

South Carolina

All counties

Tennessee

Chester
Crockett
Dyer
Fayette
Hardeman
Hardin
Haywood
Henderson
Lake
Lauderdale
Madison
McNairy
Shelby
Tipton

Climate Zone 3 Recommendation Table for Small to Medium Office Buildings

	Item	Component	Recommendation	How-To Tips	✓
Envelope	Roofs	Insulation entirely above deck	R-25.0 c.i.	EN2, 17, 19, 21–22	
		Attic and other	R-38.0	EN3, 17, 19, 20–21	
		Metal building	R-19.0 + R-10.0 FC	EN4, 17, 19, 21	
		SRI	78	EN1	
	Walls	Mass (HC > 7 Btu/ft^2)	R-11.4 c.i.	EN5, 17, 19, 21	
		Steel framed	R-13.0 + R-7.5 c.i.	EN6, 17, 19, 21	
		Wood framed and other	R-13.0 + R-3.8 c.i.	EN7, 17, 19, 21	
		Metal building	R-0.0 + R-13.0 c.i.	EN8, 17, 19, 21	
		Below-grade walls	R-7.5 c.i. (No recommendation in climate zone 3A)	EN9, 17, 19, 21–22	
	Floors	Mass	R-12.5 c.i.	EN10, 17, 19, 21	
		Steel joint	R-30.0	EN11, 17, 19, 21	
		Wood framed and other	R-30.0	EN11, 17, 19, 21	
	Slabs	Unheated	No recommendation	None	
		Heated	R-15.0 for 24 in.	EN13–14, 17, 19, 21–22	
	Doors	Swinging	U-0.70	EN15, 17–18	
		Nonswinging	U-0.50	EN16–17	
	Vestibules	At building entrance	Yes for buildings > 10,000 ft^2 only	EN18	
	Continuous Air Barriers	Continuous air barrier	Entire building envelope	EN17	
	Vertical Fenestration	WWR	20% to 40%	EN25; DL6	
		Window orientation	Area of W and E windows each less than area of S windows (N in southern hemisphere)	EN28, 30	
		Exterior sun control (S, E, and W only)	PF-0.5	EN26	
		Thermal transmittance	Nonmetal framing windows = U-0.41 Metal framing windows = U-0.60	EN23–24	
		SHGC	Nonmetal framing windows = 0.25 Metal framing windows = 0.25	EN 24, 29; DL12	
Daylighting / Lighting	Daylighting	Light-to-solar-gain ratio	Minimum VT/SHGC = 1.10	EN34–42	
		Vertical fenestration EA	0.08	DL7–11	
	Interior Finishes	Interior surface average reflectance	Ceilings = 80% Wall surfaces = 70% Open office partitions = 50%	DL14–15; EL4	
		Open office partitions parallel to window walls	Total partition height = 36 in. maximum - or - Partition above desk height = min 50% translucent	DL4, 10; EL2, 4	
	Interior Lighting	LPD	0.75 W/ft^2	DL1–5; EL2–3, 5, 12–19	
		24-hour lighting LPD	0.075 W/ft^2	EL20	
		Light source lamp efficacy (mean LPW)	T8 and T5 lamps > 2 ft = 92, T8 and T5 lamps ≤ 2 ft = 85, All other > 50	EL6–8	
		Ballasts	4 ft T8 lamp nondimming applications = NEMA Premium instant start 4 ft T8 lamp dimming applications = NEMA Premium program start Fluorescent and HID sources = electronic	EL8	
		Controls for daylight harvesting in open offices—locate on N and S sides of bldg	Dim all general fluorescent lights within primary and secondary daylight zones of open offices	DL16–20; EL1, 11	
		Automatic controls	Auto ON to 50% = private offices, conference and meeting rooms, lounge and break rooms, copy rooms, storage rooms Auto ON occupancy sensors = restrooms, electrical/ mechanical rooms, open and private office task lighting Time switch control = all other spaces	DL13; EL1, 5, 9–10	
	Exterior Lighting	Façade and landscape lighting	LPD = 0.075 W/ft^2 in LZ3 and LZ4, 0.05 W/ft^2 in LZ2 Controls = auto OFF between 12 am and 6 am	EL23	
		Parking lots and drives	LPD = 0.1 W/ft^2 in LZ3 and LZ4, 0.06 W/ft^2 in LZ2 Controls = auto reduce to 25% (12 am to 6 am)	EL21, 24–25	
		Walkways, plazas, and special feature areas	LPD = 0.16 W/ft^2 LZ3 and LZ4, 0.14 W/ft^2 in LZ2 Controls = auto reduce to 25% (12 am to 6 am)	EL22, 24–25	
		All other exterior lighting	LPD = follow Standard 90.1-2010 Controls = auto reduce to 25% (12 am to 6 am)	EL24–25	
Plug Loads	Equipment Choices	Laptop computers	Minimum 2/3 of total computers	PL1–2	
		ENERGY STAR equipment	For all computers, equipment, and appliances	PL1–2	
		Equipment power density	For all computers, equipment, and appliances	PL1–2, 4–6	
	Controls	Computer power control	Network control with power saving modes and control OFF during unoccupied hours	PL3	
		Occupancy sensors	Desk plug strip occupancy sensors	PL3	
		Timer switches	Water coolers and coffee makers control OFF during unoccupied hours	PL3	
		Vending machine control	Yes	PL3	

*Note: Where the table says "No recommendation," the user must meet the more stringent of either the applicable version of ASHRAE/IES Standard 90.1 or the local code requirements.

Climate Zone 3 Recommendation Table for Small to Medium Office Buildings *(Continued)*

	Item	Component	Recommendation	How-To Tips	✓
SWH	SWH	Gas water heater efficiency	Condensing water heaters = 90% efficiency	WH1–5	
		Electric storage EF (≤12 kW, ≥20 gal)	EF > 0.99 – 0.0012 × volume (see Table 5-7)	WH1–5	
		Point-of-use heater selection	0.81 EF or 81% E_t	WH1–5	
		Electric heat pump water heater efficiency	COP 3.0 (interior heat source)	WH1–5	
		Pipe insulation (d < 1½ in. / d ≥ 1½ in.)	1 in. / 1 ½ in.	WH6	
HVAC	Packaged Single-Zone Air-Source Heat Pumps	Cooling and heating efficiency	See Table 5-8 for efficiency	HV3	
		ESP	0.7 in. w.c.	HV3	
		DOAS air-source heat pump	Yes	HV10	
		DOAS heating and cooling efficiency	See Table 5-10 for efficiency	HV10–11	
		DOAS energy recovery	Climate zone 3A = yes, see Table 5-11 for effectiveness Climate zone 3B (CA) = No	HV12	
		DOAS fan—ESP	1.5 in. w.c. maximum	HV20	
		Unit size	5 tons or less	None	
		Single-stage cooling and heating efficiency	Cooling = 16.4 EER Heating = 5.2 COP	HV4	
		Two-stage cooling and heating efficiency	Cooling part load = 17.6 EER Cooling full load = 15.0 EER Heating part load = 5.7 COP Heating full load = 5.0 COP	HV4	
	WSHPs with DOAS	WSHP fan—ESP	0.5 in. w.c.	HV4, 20	
		Condensing boiler efficiency	90%	HV14, 30	
		DOAS water-to-water heat pump	See Table 5-10 for efficiency	HV10–11	
		DOAS variable airflow with DCV	Yes	HV17–18	
		DOAS energy recovery	Yes, see Table 5-11 for effectiveness	HV12	
		DOAS fan and motor	65% mechanical efficiency Motor efficiency per Standard 90.1-2010, Table 10.8B VSD efficiency = 95%	HV23	
	VAV DX with Indirect Gas-Fired or Elextric Internal Heat and Electric Perimeter Heat	DX efficiency	See Table 5-9 for efficiency	HV6	
		Low-temperature air supply and SAT reset	50°F to 58°F	HV6, 11, 25, 31	
		Perimeter convector heat source	Electric	HV6	
		Gas furnace in DX units	No recommendation	HV6	
		Economizer	≥54,000 Btu/h, differential enthalpy control (Climate zone 3A) ≥54,000 Btu/h, differential dry-bulb control (Climate zones 3B and 3C)	HV6, 16	
		Energy recovery	Yes, see Table 5-11 for effectiveness	HV12	
		Indirect evaporative cooling	Climate zone 3B only	HV13, 36	
		Demand control and ventilation reset	Yes	HV17–18	
		ESP	2.0 in. w.c.	HV7, 20	
	VAV CHW (same as VAV DX except...)	Air-cooled chiller efficiency	10 EER	HV14, 29	
		Air-cooled chiller IPLV	12.5 IPLV < 150 tons, 12.75 IPLV ≥ 150 tons	HV14	
		Variable-speed pumping	Yes	HV29	
		Maximum fan power	0.72 W/cfm	HV7, 23, 31	
	Fan-Coils with DOAS	Air-cooled chiller efficiency	10 EER	HV14, 29	
		Air-cooled chiller IPLV	12.5 IPLV < 150 tons 12.75 IPLV ≥ 150 tons	HV14	
		Condensing boiler efficiency	90%	HV14, 30	
		Variable-speed pumping	Yes	HV29–30	
		VAV fan-coil units	Yes	HV8	
		Fan-coil unit fan power	0.30 W/cfm	HV8	
		DOAS chilled-water and hot-water coils served by same plant as fan-coils	Yes	HV10	
		DOAS variable airflow with DCV	Yes	HV10–11, 17	
		DOAS energy recovery	Yes, see Table 5-11 for effectiveness	HV12	
		DOAS fan and motor	65% mechanical efficiency, motor efficiency Standard 90.1-2010, Table 10.8B	HV23	
	Radiant Systems with DOAS	Air-cooled chiller full-load efficiency	10 EER	HV14, 29	
		Air-cooled chiller IPLV	12.5 IPLV < 150 tons, 12.75 IPLV ≥ 150 tons	HV14	
		Condensing boiler efficiency	90%	HV14, 30	
		DOAS heating and cooling efficiency	See Table 5-10 for efficiency	HV10–11	
		DOAS variable airflow with DCV	Yes	HV10–11, 17–18	
		DOAS energy recovery	Yes, see Table 5-11 for effectiveness	HV12	
		DOAS fan—ESP	1.5 in. w.c.	HV20	
	Ducts and Dampers	OA damper	Motorized damper	HV16	
		Friction rate	0.08 in./100 ft	HV20	
		Sealing	Seal Class B	HV22	
		Location	Interior only	HV20	
		Insulation level	R-6.0	HV21	

*Note: Where the table says "No recommendation," the user must meet the more stringent of either the applicable version of ASHRAE/IES Standard 90.1 or the local code requirements.

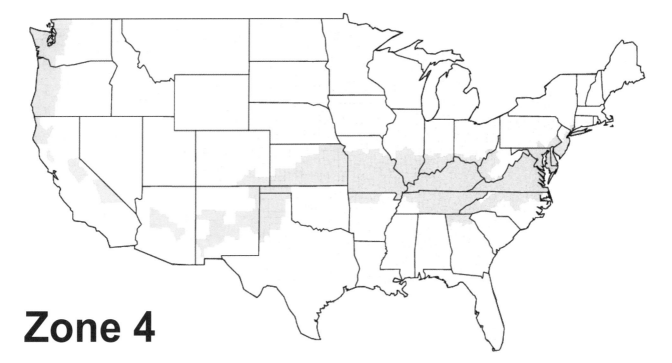

Zone 4

Arizona
Gila
Yavapai

Arkansas
Baxter
Benton
Boone
Carroll
Fulton
Izard
Madison
Marion
Newton
Searcy
Stone
Washington

California
Amador
Calaveras
Del Norte
El Dorado
Humboldt
Inyo
Lake
Mariposa
Trinity
Tuolumne

Colorado
Baca
Las Animas
Otero

Delaware
All counties

District of Columbia

Georgia
Banks
Catoosa
Chattooga
Dade
Dawson
Fannin
Floyd
Franklin
Gilmer
Gordon
Habersham
Hall
Lumpkin
Murray
Pickens
Rabun
Stephens
Towns
Union

Walker
White
Whitfield

Illinois
Alexander
Bond
Brown
Christian
Clay
Clinton
Crawford
Edwards
Effingham
Fayette
Franklin
Gallatin
Hamilton
Hardin
Jackson
Jasper
Jefferson
Johnson
Lawrence
Macoupin
Madison
Marion
Massac
Monroe
Montgomery
Perry
Pope
Pulaski
Randolph
Richland
Saline
Shelby
St. Clair
Union
Wabash
Washington
Wayne
White
Williamson

Indiana
Clark
Crawford
Daviess
Dearborn
Dubois
Floyd
Gibson
Greene
Harrison
Jackson
Jefferson
Jennings
Knox
Lawrence
Martin
Monroe

Ohio
Orange
Perry
Pike
Posey
Ripley
Scott
Spencer
Sullivan
Switzerland
Vanderburgh
Warrick
Washington

Kansas
All counties except:
Cheyenne
Cloud
Decatur
Ellis
Gove
Graham
Greeley
Hamilton
Jewell
Lane
Logan
Mitchell
Ness
Norton
Osborne
Phillips
Rawlins
Republic
Rooks
Scott
Sheridan
Sherman
Smith
Thomas
Trego
Wallace
Wichita

Kentucky
All counties

Maryland
All counties except:
Garrett

Missouri
All counties except:
Adair
Andrew
Atchison
Buchanan
Caldwell
Chariton
Clark
Clinton
Daviess

DeKalb
Gentry
Grundy
Harrison
Holt
Knox
Lewis
Linn
Livingston
Macon
Marion
Mercer
Nodaway
Pike
Putnam
Ralls
Schuyler
Scotland
Shelby
Sullivan
Worth

New Jersey
All counties except:
Bergen
Hunterdon
Mercer
Morris
Passaic
Somerset
Sussex
Warren

New Mexico
Bernalillo
Cibola
Curry
DeBaca
Grant
Guadalupe
Lincoln
Quay
Roosevelt
Sierra
Socorro
Union
Valencia

New York
Bronx
Kings
Nassau
New York
Queens
Richmond
Suffolk
Westchester

North Carolina
Alamance
Alexander

Bertie
Buncombe
Burke
Caldwell
Caswell
Catawba
Chatham
Cherokee
Clay
Cleveland
Davie
Durham
Forsyth
Franklin
Gates
Graham
Granville
Guilford
Halifax
Harnett
Haywood
Henderson
Hertford
Iredell
Jackson
Lee
Lincoln
Macon
Madison
McDowell
Nash
Northampton
Orange
Person
Polk
Rockingham
Rutherford
Stokes
Surry
Swain
Transylvania
Vance
Wake
Warren
Wilkes
Yadkin

Ohio
Adams
Brown
Clermont
Gallia
Hamilton
Lawrence
Pike
Scioto
Washington

Oklahoma
Beaver
Cimarron
Texas

Oregon
Benton
Clackamas
Clatsop
Columbia
Coos
Curry
Douglas
Jackson
Josephine
Lane
Lincoln
Linn
Marion
Multnomah
Polk
Tillamook
Washington
Yamhill

Pennsylvania
Bucks
Chester
Delaware
Montgomery
Philadelphia
York

Tennessee
All counties except:
Chester
Crockett
Dyer
Fayette
Hardeman
Hardin
Haywood
Henderson
Lake
Lauderdale
Madison
McNairy
Shelby
Tipton

Texas
Armstrong
Bailey
Briscoe
Carson
Castro
Cochran
Dallam
Deaf Smith
Donley
Floyd
Gray
Hale
Hansford
Hartley
Hockley
Hutchinson
Lamb

Lipscomb
Moore
Ochiltree
Oldham
Parmer
Potter
Randall
Roberts
Sherman
Swisher
Yoakum

Virginia
All counties

Washington
Clallam
Clark
Cowlitz
Grays Harbor
Island
Jefferson
King
Kitsap
Lewis
Mason
Pacific
Pierce
San Juan
Skagit
Snohomish
Thurston
Wahkiakum
Whatcom

West Virginia
Berkeley
Boone
Braxton
Cabell
Calhoun
Clay
Gilmer
Jackson
Jefferson
Kanawha
Lincoln
Logan
Mason
McDowell
Mercer
Mingo
Monroe
Morgan
Pleasants
Putnam
Ritchie
Roane
Tyler
Wayne
Wirt
Wood
Wyoming

Climate Zone 4 Recommendation Table for Small to Medium Office Buildings

	Item	Component	Recommendation	How-To Tips	✓
Envelope	Roofs	Insulation entirely above deck	R-30.0 c.i.	EN2, 17, 19, 21–22	
		Attic and other	R-49.0	EN3, 17, 19, 20–21	
		Metal building	R-19.0 + R-11.0 Ls	EN4, 17, 19, 21	
		SRI	No recommendation	None	
	Walls	Mass (HC > 7 Btu/ft^2)	R-13.3 c.i.	EN5, 17, 19, 21	
		Steel framed	R-13.0 + R-7.5 c.i.	EN6, 17, 19, 21	
		Wood framed and other	R-13.0 + R-7.5 c.i.	EN7, 17, 19, 21	
		Metal building	R-0.0 + R-15.8 c.i.	EN8, 17, 19, 21	
		Below-grade walls	R-7.5 c.i.	EN9, 17, 19, 21–22	
	Floors	Mass	R-14.6 c.i.	EN10, 17, 19, 21	
		Steel joint	R-38.0	EN11, 17, 19, 21	
		Wood framed and other	R-38.0	EN11, 17, 19, 21	
	Slabs	Unheated	R-15.0 for 24 in.	EN12, 14, 17, 19, 21–22	
		Heated	R-20.0 for 24 in.	EN13–14, 17, 19, 21–22	
	Doors	Swinging	U-0.50	EN15, 17–18	
		Nonswinging	U-0.50	EN16–17	
	Vestibules	At building entrance	Yes	EN18	
	Continuous Air Barriers	Continuous air barrier	Entire building envelope	EN17	
	Vertical Fenestration	WWR	20% to 40%	EN25; DL6	
		Window orientation	Area of W and E windows each less than area of S windows (N in southern hemisphere)	EN30–31	
		Exterior sun control (S, E, and W only)	PF-0.5	EN26	
		Thermal transmittance	Nonmetal framing windows = U-0.38 Metal framing windows = U-0.39	EN23–24	
		SHGC	Nonmetal framing windows = 0.26 Metal framing windows = 0.38	EN24, 32–33; DL12	
Daylighting / Lighting	Daylighting	Light-to-solar-gain ratio	Minimum VT/SHGC = 1.10	EN34–42	
		Vertical fenestration EA	0.12	DL7–11	
	Interior Finishes	Interior surface average reflectance	Ceilings = 80% Wall surfaces = 70% Open office partitions = 50%	DL14–15; EL4	
		Open office partitions parallel to window walls	Total partition height = 36 in. maximum - or - Partition above desk height = min 50% translucent	DL4, 10; EL2, 4	
	Interior Lighting	LPD	0.75 W/ft^2	DL1–5; EL2–3, 5, 12–19	
		24-hour lighting LPD	0.075 W/ft^2	EL20	
		Light source lamp efficacy (mean LPW)	T8 and T5 lamps > 2 ft = 92 T8 and T5 lamps ≤ 2 ft = 85 All other > 50	EL6–8	
		Ballasts	4 ft T8 lamp nondimming applications = NEMA Premium instant start 4 ft T8 lamp dimming applications = NEMA Premium program start Fluorescent and HID sources = electronic	EL8	
		Controls for daylight harvesting in open offices—locate on N and S sides of bldg	Dim all general fluorescent lights within primary and secondary daylight zones of open offices	DL16–20; EL1, 11	
		Automatic controls	Auto ON to 50% = private offices, conference and meeting rooms, lounge and break rooms, copy rooms, storage rooms Auto ON occupancy sensors = restrooms, electrical/ mechanical rooms, open and private office task lighting Time switch control = all other spaces	DL13; EL1, 5, 9–10	
	Exterior Lighting	Façade and landscape lighting	LPD = 0.075 W/ft^2 in LZ3 and LZ4, 0.05 W/ft^2 in LZ2 Controls = auto OFF between 12 am and 6 am	EL23	
		Parking lots and drives	LPD = 0.1 W/ft^2 in LZ3 and LZ4, 0.06 W/ft^2 in LZ2 Controls = auto reduce to 25% (12 am to 6 am)	EL21, 24–25	
		Walkways, plazas, and special feature areas	LPD = 0.16 W/ft^2 LZ3 and LZ4, 0.14 W/ft^2 in LZ2 Controls = auto reduce to 25% (12 am to 6 am)	EL22, 24–25	
		All other exterior lighting	LPD = follow Standard 90.1-2010 Controls = auto reduce to 25% (12 am to 6 am)	EL24–25	
Plug Loads	Equipment Choices	Laptop computers	Minimum 2/3 of total computers	PL1–2	
		ENERGY STAR equipment	For all computers, equipment, and appliances	PL1–2	
		Equipment power density	For all computers, equipment, and appliances	PL1–2, 4–6	
	Controls	Computer power control	Network control with power saving modes and control OFF during unoccupied hours	PL3	
		Occupancy sensors	Desk plug strip occupancy sensors	PL3	
		Timer switches	Water coolers and coffee makers control OFF during unoccupied hours	PL3	
		Vending machine control	Yes	PL3	

*Note: Where the table says "No recommendation," the user must meet the more stringent of either the applicable version of ASHRAE/IES Standard 90.1 or the local code requirements.

Climate Zone 4 Recommendation Table for Small to Medium Office Buildings *(Continued)*

	Item	Component	Recommendation	How-To Tips	✓
SWH	SWH	Gas water heater efficiency	Condensing water heaters = 90% efficiency	WH1–5	
		Electric storage EF (≤12 kW, ≥20 gal)	EF > 0.99 − 0.0012 × volume (see Table 5-7)	WH1–5	
		Point-of-use heater selection	0.81 EF or 81% E_t	WH1–5	
		Electric heat pump water heater efficiency	COP 3.0 (interior heat source)	WH1–5	
		Pipe insulation (d < 1½ in. / d ≥ 1½ in.)	1 in. / 1 ½ in.	WH6	
HVAC	Packaged Single-Zone Air-Source Heat Pumps	Cooling and heating efficiency	See Table 5-8 for efficiency	HV3	
		ESP	0.7 in. w.c.	HV3	
		DOAS air-source heat pump	Yes	HV10	
		DOAS heating and cooling efficiency	See Table 5-10 for efficiency	HV10–11	
		DOAS energy recovery	Yes, see Table 5-11 for effectiveness	HV12	
		DOAS fan—ESP	1.5 in. w.c. maximum	HV20	
	WSHPs with DOAS	Unit size	5 tons or less	None	
		Single-stage cooling and heating efficiency	Cooling = 16.4 EER Heating = 5.2 COP	HV4	
		Two-stage cooling and heating efficiency	Cooling part load = 17.6 EER Cooling full load = 15.0 EER Heating part load = 5.7 COP Heating full load = 5.0 COP	HV4	
		WSHP fan—ESP	0.5 in. w.c.	HV4, 20	
		Condensing boiler efficiency	90%	HV14, 30	
		DOAS water-to-water heat pump	See Table 5-10 for efficiency	HV10–11	
		DOAS variable airflow with DCV	Yes	HV17–18	
		DOAS energy recovery	Yes, see Table 5-11 for effectiveness	HV12	
		DOAS fan and motor	65% mechanical efficiency Motor efficiency per Standard 90.1-2010, Table 10.8B VSD efficiency = 95%	HV23	
	VAV DX with Indirect Gas-Fired or Elextric Internal Heat and Electric Perimeter Heat	DX efficiency	See Table 5-9 for efficiency	HV6	
		Low-temperature air supply and SAT reset	50°F to 58°F	HV6, 11, 25, 31	
		Perimeter convector heat source	Electric	HV6	
		Gas furnace in DX units	No recommendation	HV6	
		Economizer	≥54,000 Btu/h, differential enthalpy control (Climate zone 4A) ≥54,000 Btu/h, differential dry-bulb control (Climate zones 4B and 4C)	HV6, 16	
		Energy recovery	Yes, see Table 5-11 for effectiveness	HV12	
		Indirect evaporative cooling	Climate zone 4B only	HV13, 36	
		Demand control and ventilation reset	Yes	HV17–18	
		ESP	2.0 in. w.c.	HV7, 20	
	VAV CHW (same as VAV DX except...)	Air-cooled chiller efficiency	10 EER	HV14, 29	
		Air-cooled chiller IPLV	12.5 IPLV < 150 tons, 12.75 IPLV ≥ 150 tons	HV14	
		Variable-speed pumping	Yes	HV29	
		Maximum fan power	0.72 W/cfm	HV7, 23, 31	
	Fan-Coils with DOAS	Air-cooled chiller efficiency	10 EER	HV14, 29	
		Air-cooled chiller IPLV	12.5 IPLV < 150 tons 12.75 IPLV ≥ 150 tons	HV14	
		Condensing boiler efficiency	90%	HV14, 30	
		Variable-speed pumping	Yes	HV29–30	
		VAV fan-coil units	Yes	HV8	
		Fan-coil unit fan power	0.30 W/cfm	HV8	
		DOAS chilled-water and hot-water coils served by same plant as fan-coils	Yes	HV10	
		DOAS variable airflow with DCV	Yes	HV10–11, 17	
		DOAS energy recovery	Yes, see Table 5-11 for effectiveness	HV12	
		DOAS fan and motor	65% mechanical efficiency, motor efficiency Standard 90.1-2010, Table 10.8B	HV23	
	Radiant Systems with DOAS	Air-cooled chiller full-load efficiency	10 EER	HV14, 29	
		Air-cooled chiller IPLV	12.5 IPLV < 150 tons, 12.75 IPLV ≥ 150 tons	HV14	
		Condensing boiler efficiency	90%	HV14, 30	
		DOAS heating and cooling efficiency	See Table 5-10 for efficiency	HV10–11	
		DOAS variable airflow with DCV	Yes	HV10–11, 17–18	
		DOAS energy recovery	Yes, see Table 5-11 for effectiveness	HV12	
		DOAS fan—ESP	1.5 in. w.c.	HV20	
	Ducts and Dampers	OA damper	Motorized damper	HV16	
		Friction rate	0.08 in./100 ft	HV20	
		Sealing	Seal Class B	HV22	
		Location	Interior only	HV20	
		Insulation level	R-6.0	HV21	

*Note: Where the table says "No recommendation," the user must meet the more stringent of either the applicable version of ASHRAE/IES Standard 90.1 or the local code requirements.

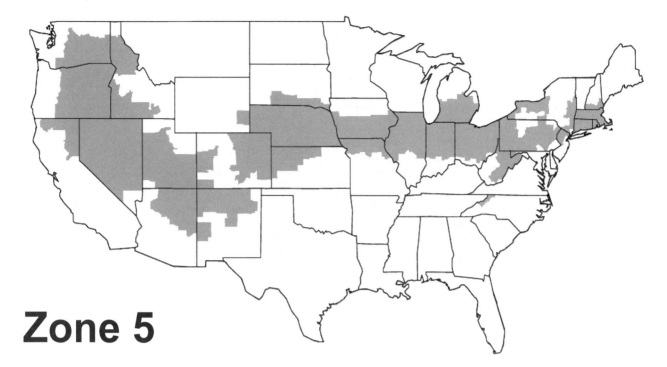

Zone 5

Arizona
Apache
Coconino
Navajo

California
Lassen
Modoc
Nevada
Plumas
Sierra
Siskiyou

Colorado
Adams
Arapahoe
Bent
Boulder
Cheyenne
Crowley
Delta
Denver
Douglas
Elbert
El Paso
Fremont
Garfield
Gilpin
Huerfano
Jefferson
Kiowa
Kit Carson
La Plata
Larimer
Lincoln
Logan
Mesa
Montezuma
Montrose
Morgan
Phillips
Prowers
Pueblo
Sedgwick
Teller
Washington
Weld
Yuma

Connecticut
All counties

Idaho
Ada
Benewah
Canyon
Cassia
Clearwater
Elmore
Gem
Gooding
Idaho
Jerome
Kootenai
Latah
Lewis
Lincoln
Minidoka
Nez Perce
Owyhee
Payette
Power
Shoshone

Twin Falls
Washington

Illinois
*All counties
except:*
Alexander
Bond
Christian
Clay
Clinton
Crawford
Edwards
Effingham
Fayette
Franklin
Gallatin
Hamilton
Hardin
Jackson
Jasper
Jefferson
Johnson
Lawrence
Macoupin
Madison
Marion
Massac
Monroe
Montgomery
Perry
Pope
Pulaski
Randolph
Richland
Saline
Shelby
St. Clair
Union
Wabash
Washington
Wayne
White
Williamson
Brown

Indiana
*All counties
except:*
Clark
Crawford
Daviess
Dearborn
Dubois
Floyd
Gibson
Greene
Harrison
Jackson
Jefferson
Jennings
Knox
Lawrence
Martin
Monroe
Ohio
Orange
Perry
Pike
Posey
Ripley
Scott
Spencer
Sullivan

Switzerland
Vanderburgh
Warrick
Washington

Iowa
*All counties
except:*
Allamakee
Black Hawk
Bremer
Buchanan
Buena Vista
Butler
Calhoun
Cerro Gordo
Cherokee
Chickasaw
Clay
Clayton
Delaware
Dickinson
Emmet
Fayette
Floyd
Franklin
Grundy
Hamilton
Hancock
Hardin
Howard
Humboldt
Ida
Kossuth
Lyon
Mitchell
O'Brien
Osceola
Palo Alto
Plymouth
Pocahontas
Sac
Sioux
Webster
Winnebago
Winneshiek
Worth
Wright

Kansas
Cheyenne
Cloud
Decatur
Ellis
Gove
Graham
Greeley
Hamilton
Jewell
Lane
Logan
Mitchell
Ness
Norton
Osborne
Phillips
Rawlins
Republic
Rooks
Scott
Sheridan
Sherman
Smith
Thomas

Trego
Wallace
Wichita

Maryland
Garrett

Massachusetts
All counties

Michigan
Allegan
Barry
Bay
Berrien
Branch
Calhoun
Cass
Clinton
Eaton
Genesee
Gratiot
Hillsdale
Ingham
Ionia
Jackson
Kalamazoo
Kent
Lapeer
Lenawee
Livingston
Macomb
Midland
Monroe
Montcalm
Muskegon
Oakland
Ottawa
Saginaw
Shiawassee
St. Clair
St. Joseph
Tuscola
Van Buren
Washtenaw
Wayne

Missouri
Adair
Andrew
Atchison
Buchanan
Caldwell
Chariton
Clark
Clinton
Daviess
DeKalb
Gentry
Grundy
Harrison
Holt
Knox
Lewis
Linn
Livingston
Macon
Marion
Mercer
Nodaway
Pike
Putnam
Ralls
Schuyler

Scotland
Shelby
Sullivan
Worth

Nebraska
All counties

Nevada
*All counties
except:*
Clark

New Hampshire
Cheshire
Hillsborough
Rockingham
Strafford

New Jersey
Bergen
Hunterdon
Mercer
Morris
Passaic
Somerset
Sussex
Warren

New Mexico
Catron
Colfax
Harding
Los Alamos
McKinley
Mora
Rio Arriba
Sandoval
San Juan
San Miguel
Santa Fe
Taos
Torrance

New York
Albany
Cayuga
Chautauqua
Chemung
Columbia
Cortland
Dutchess
Erie
Genesee
Greene
Livingston
Monroe
Niagara
Onondaga
Ontario
Orange
Orleans
Oswego
Putnam
Rensselaer
Rockland
Saratoga
Schenectady
Seneca
Tioga
Washington
Wayne
Yates

North Carolina
Alleghany
Ashe
Avery
Mitchell
Watauga
Yancey

Ohio
*All counties
except:*
Adams
Brown
Clermont
Gallia
Hamilton
Lawrence
Pike
Scioto
Washington

Oregon
Baker
Crook
Deschutes
Gilliam
Grant
Harney
Hood River
Jefferson
Klamath
Lake
Malheur
Morrow
Sherman
Umatilla
Union
Wallowa
Wasco
Wheeler

Pennsylvania
*All counties
except:*
Bucks
Cameron
Chester
Clearfield
Delaware
Elk
McKean
Montgomery
Philadelphia
Potter
Susquehanna
Tioga
Wayne
York

Rhode Island
All counties

South Dakota
Bennett
Bon Homme
Charles Mix
Clay
Douglas
Gregory
Hutchinson
Jackson
Mellette
Todd

Tripp
Union
Yankton

Utah
*All counties
except:*
Box Elder
Cache
Carbon
Daggett
Duchesne
Morgan
Rich
Summit
Uintah
Wasatch
Washington

Washington
Adams
Asotin
Benton
Chelan
Columbia
Douglas
Franklin
Garfield
Grant
Kittitas
Klickitat
Lincoln
Skamania
Spokane
Walla Walla
Whitman
Yakima

Wyoming
Goshen
Platte

West Virginia
Barbour
Brooke
Doddridge
Fayette
Grant
Greenbrier
Hampshire
Hancock
Hardy
Harrison
Lewis
Marion
Marshall
Mineral
Monongalia
Nicholas
Ohio
Pendleton
Pocahontas
Preston
Raleigh
Randolph
Summers
Taylor
Tucker
Upshur
Webster
Wetzel

Climate Zone 5 Recommendation Table for Small to Medium Office Buildings

	Item	Component	Recommendation	How-To Tips	✓
Envelope	Roofs	Insulation entirely above deck	R-30.0 c.i.	EN2, 17, 19, 21–22	
		Attic and other	R-49.0	EN3, 17, 19, 20–21	
		Metal building	R-19.0 + R-11.0 Ls	EN4, 17, 19, 21	
		SRI	No recommendation	None	
	Walls	Mass (HC > 7 Btu/ft^2)	R-13.3 c.i.	EN5, 17, 19, 21	
		Steel framed	R-13.0 + R-15.6 c.i.	EN6, 17, 19, 21	
		Wood framed and other	R-13.0 + R-10.0 c.i.	EN7, 17, 19, 21	
		Metal building	R-0.0 + R-19.0 c.i.	EN8, 17, 19, 21	
		Below-grade walls	R-7.5 c.i.	EN9, 17, 19, 21–22	
	Floors	Mass	R-14.6 c.i.	EN10, 17, 19, 21	
		Steel joint	R-38.0	EN11, 17, 19, 21	
		Wood framed and other	R-38.0	EN11, 17, 19, 21	
	Slabs	Unheated	R-15.0 for 24 in.	EN12, 14, 17, 19, 21–22	
		Heated	R-20.0 for 24 in.	EN13–14, 17, 19, 21–22	
	Doors	Swinging	U-0.50	EN15, 17–18	
		Nonswinging	U-0.50	EN16–17	
	Vestibules	At building entrance	Yes	EN18	
	Continuous Air Barriers	Continuous air barrier	Entire building envelope	EN17	
	Vertical Fenestration	WWR	20% to 40%	EN25; DL6	
		Window orientation	Area of W and E windows each less than area of S windows (N in southern hemisphere)	EN30–31	
		Exterior sun control (S, E, and W only)	PF-0.5	EN26	
		Thermal transmittance	Nonmetal framing windows = U-0.35 Metal framing windows = U-0.39	EN23–24	
		SHGC	Nonmetal framing windows = 0.26 Metal framing windows = 0.38	EN24, 32–33; DL12	
		Light-to-solar-gain ratio	Minimum VT/SHGC = 1.10	EN34–42	
Daylighting / Lighting	Daylighting	Vertical fenestration EA	0.12	DL7–11	
	Interior Finishes	Interior surface average reflectance	Ceilings = 80% Wall surfaces = 70% Open office partitions = 50%	DL14–15; EL4	
		Open office partitions parallel to window walls	Total partition height = 36 in. maximum - or - Partition above desk height = min 50% translucent	DL4, 10; EL2, 4	
	Interior Lighting	LPD	0.75 W/ft^2	DL1–5; EL2–3, 5, 12–19	
		24-hour lighting LPD	0.075 W/ft^2	EL20	
		Light source lamp efficacy (mean LPW)	T8 and T5 lamps > 2 ft = 92 T8 and T5 lamps ≤ 2 ft = 85 All other > 50	EL6–8	
		Ballasts	4 ft T8 lamp nondimming applications = NEMA Premium instant start 4 ft T8 lamp dimming applications = NEMA Premium program start Fluorescent and HID sources = electronic	EL8	
		Controls for daylight harvesting in open offices—locate on N and S sides of bldg	Dim all general fluorescent lights within primary and secondary daylight zones of open offices	DL16–20; EL1, 11	
		Automatic controls	Auto ON to 50% = private offices, conference and meeting rooms, lounge and break rooms, copy rooms, storage rooms Auto ON occupancy sensors = restrooms, electrical/ mechanical rooms, open and private office task lighting Time switch control = all other spaces	DL13; EL1, 5, 9–10	
	Exterior Lighting	Façade and landscape lighting	LPD = 0.075 W/ft^2 in LZ3 and LZ4, 0.05 W/ft^2 in LZ2 Controls = auto OFF between 12 am and 6 am	EL23	
		Parking lots and drives	LPD = 0.1 W/ft^2 in LZ3 and LZ4, 0.06 W/ft^2 in LZ2 Controls = auto reduce to 25% (12 am to 6 am)	EL21, 24–25	
		Walkways, plazas, and special feature areas	LPD = 0.16 W/ft^2 LZ3 and LZ4, 0.14 W/ft^2 in LZ2 Controls = auto reduce to 25% (12 am to 6 am)	EL22, 24–25	
		All other exterior lighting	LPD = follow Standard 90.1-2010 Controls = auto reduce to 25% (12 am to 6 am)	EL24–25	
Plug Loads	Equipment Choices	Laptop computers	Minimum 2/3 of total computers	PL1–2	
		ENERGY STAR equipment	For all computers, equipment, and appliances	PL1–2	
		Equipment power density	For all computers, equipment, and appliances	PL1–2, 4–6	
	Controls	Computer power control	Network control with power saving modes and control OFF during unoccupied hours	PL3	
		Occupancy sensors	Desk plug strip occupancy sensors	PL3	
		Timer switches	Water coolers and coffee makers control OFF during unoccupied hours	PL3	
		Vending machine control	Yes	WH4	

*Note: Where the table says "No recommendation," the user must meet the more stringent of either the applicable version of ASHRAE/IES Standard 90.1 or the local code requirements.

Climate Zone 5 Recommendation Table for Small to Medium Office Buildings *(Continued)*

	Item	Component	Recommendation	How-To Tips	✓
SWH	SWH	Gas water heater efficiency	Condensing water heaters = 90% efficiency	WH1–5	
		Electric storage EF (≤12 kW, ≥20 gal)	EF > 0.99 − 0.0012 × volume (see Table 5-7)	WH1–5	
		Point-of-use heater selection	0.81 EF or 81% E_t	WH1–5	
		Electric heat pump water heater efficiency	COP 3.0 (interior heat source)	WH1–5	
		Pipe insulation (d < 1½ in. / d ≥ 1½ in.)	1 in. / 1 ½ in.	WH4	
HVAC	Packaged Single-Zone Air-Source Heat Pumps	Cooling and heating efficiency	See Table 5-8 for efficiency	HV3	
		ESP	0.7 in. w.c.	HV3	
		DOAS air-source heat pump	Yes	HV10	
		DOAS heating and cooling efficiency	See Table 5-10 for efficiency	HV10–11	
		DOAS energy recovery	Yes, see Table 5-11 for effectiveness	HV12	
		DOAS fan—ESP	1.5 in. w.c. maximum	HV20	
		Unit size	5 tons or less	None	
	WSHPs with DOAS	Single-stage cooling and heating efficiency	Cooling = 16.4 EER Heating = 5.2 COP	HV4	
		Two-stage cooling and heating efficiency	Cooling part load = 17.6 EER Cooling full load = 15.0 EER Heating part load = 5.7 COP Heating full load = 5.0 COP	HV4	
		WSHP fan—ESP	0.5 in. w.c.	HV4, 20	
		Condensing boiler efficiency	90%	HV14, 30	
		DOAS water-to-water heat pump	See Table 5-10 for efficiency	HV10–11	
		DOAS variable airflow with DCV	Yes	HV17–18	
		DOAS energy recovery	Yes, see Table 5-11 for effectiveness	HV12	
		DOAS fan and motor	65% mechanical efficiency Motor efficiency per Standard 90.1-2010, Table 10.8B VSD efficiency = 95%	HV23	
	VAV DX with Gas-Fired Hydronic Heating	DX efficiency	See Table 5-9 for efficiency	HV6	
		Low-temperature air supply and SAT reset	50°F to 61°F	HV6, 11, 25, 31	
		Perimeter convector heat source	Hot water	None	
		Condensing boiler efficiency	90%	HV14, 30	
		Economizer	≥54,000 Btu/h, differential dry-bulb control	HV6, 16	
		Energy recovery	Yes, see Table 5-11 for effectiveness	HV12	
		Indirect evaporative cooling	Climate zone 5B only	HV13, 36	
		Demand control and ventilation reset	Yes	HV17–18	
		ESP	2.0 in. w.c.	HV7, 20	
	VAV CHW (same as VAV DX except...)	Air-cooled chiller efficiency	10 EER	HV14, 29	
		Air-cooled chiller IPLV	12.5 IPLV < 150 tons, 12.75 IPLV ≥ 150 tons	HV14	
		Variable-speed pumping	Yes	HV29	
		Maximum fan power	0.72 W/cfm	HV7, 23, 31	
		Air-cooled chiller efficiency	10 EER	HV14, 29	
		Air-cooled chiller IPLV	12.5 IPLV < 150 tons 12.75 IPLV ≥ 150 tons	HV14	
		Condensing boiler efficiency	90%	HV14, 30	
		Variable-speed pumping	Yes	HV29–30	
	Fan-Coils with DOAS	VAV fan-coil units	Yes	HV8	
		Fan-coil unit fan power	0.30 W/cfm	HV8	
		DOAS chilled-water and hot-water coils served by same plant as fan-coils	Yes	HV10	
		DOAS variable airflow with DCV	Yes	HV10–11, 17	
		DOAS energy recovery	Yes, see Table 5-11 for effectiveness	HV12	
		DOAS fan and motor	65% mechanical efficiency, motor efficiency Standard 90.1-2010, Table 10.8B	HV23	
	Radiant Systems with DOAS	Air-cooled chiller full-load efficiency	10 EER	HV14	
		Air-cooled chiller IPLV	12.5 IPLV < 150 tons, 12.75 IPLV ≥ 150 tons	HV14	
		Condensing boiler efficiency	90%	HV14, 30	
		DOAS heating and cooling efficiency	See Table 5-10 for efficiency	HV10–11	
		DOAS variable airflow with DCV	Yes	HV10–11, 17–18	
		DOAS energy recovery	Yes, see Table 5-11 for effectiveness	HV12	
		DOAS fan—ESP	1.5 in. w.c.	HV20	
	Ducts and Dampers	OA damper	Motorized damper	HV16	
		Friction rate	0.08 in./100 ft	HV20	
		Sealing	Seal Class B	HV22	
		Location	Interior only	HV20	
		Insulation level	R-6.0	HV21	

*Note: Where the table says "No recommendation," the user must meet the more stringent of either the applicable version of ASHRAE/IES Standard 90.1 or the local code requirements.

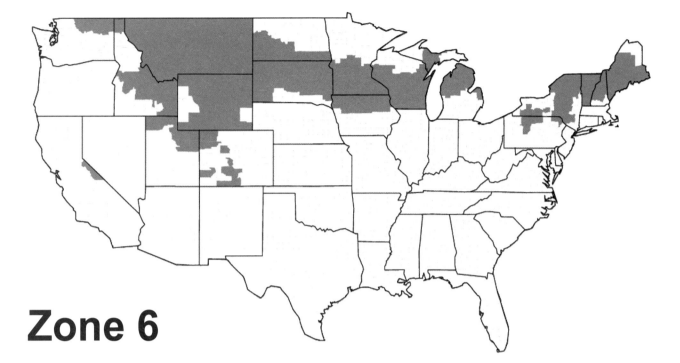

Zone 6

California

Alpine
Mono

Colorado

Alamosa
Archuleta
Chaffee
Conejos
Costilla
Custer
Dolores
Eagle
Moffat
Ouray
Rio Blanco
Saguache
San Miguel

Idaho

Adams
Bannock
Bear Lake
Bingham
Blaine
Boise
Bonner
Bonneville
Boundary
Butte
Camas
Caribou
Clark
Custer
Franklin
Fremont
Jefferson
Lemhi
Madison
Oneida
Teton
Valley

Iowa

Allamakee
Black Hawk
Bremer
Buchanan
Buena Vista
Butler
Calhoun
Cerro Gordo
Cherokee
Chickasaw
Clay
Clayton
Delaware
Dickinson
Emmet
Fayette
Floyd
Franklin
Grundy
Hamilton
Hancock
Hardin
Howard
Humboldt
Ida
Kossuth
Lyon
Mitchell
O'Brien
Osceola
Palo Alto
Plymouth
Pocahontas
Sac
Sioux
Webster
Winnebago
Winneshiek
Worth
Wright

Maine

All counties except:
Aroostook

Michigan

Alcona
Alger
Alpena
Antrim
Arenac
Benzie
Charlevoix
Cheboygan
Clare
Crawford
Delta
Dickinson
Emmet
Gladwin
Grand Traverse
Huron
Iosco
Isabella
Kalkaska
Lake
Leelanau
Manistee
Marquette
Mason
Mecosta
Menominee
Missaukee
Montmorency
Newaygo
Oceana
Ogemaw
Osceola
Oscoda
Otsego
Presque Isle
Roscommon
Sanilac
Wexford

Minnesota

Anoka
Benton
Big Stone
Blue Earth
Brown
Carver
Chippewa

Chisago
Cottonwood
Dakota
Dodge
Douglas
Faribault
Fillmore
Freeborn
Goodhue
Hennepin
Houston
Isanti
Jackson
Kandiyohi
Lac qui Parle
Le Sueur
Lincoln
Lyon
Martin
McLeod
Meeker
Morrison
Mower
Murray
Nicollet
Nobles
Olmsted
Pipestone
Pope
Ramsey
Redwood
Renville
Rice
Rock
Scott
Sherburne
Sibley
Stearns
Steele
Stevens
Swift
Todd
Traverse
Wabasha
Waseca
Washington
Watonwan
Winona
Wright
Yellow Medicine

Montana

All counties

New Hampshire

Belknap
Carroll
Coos
Grafton
Merrimack
Sullivan

New York

Allegany
Broome
Cattaraugus
Chenango
Clinton
Delaware
Essex
Franklin
Fulton
Hamilton
Herkimer
Jefferson
Lewis
Madison
Montgomery
Oneida
Otsego
Schoharie
Schuyler
Steuben
St. Lawrence
Sullivan
Tompkins
Ulster
Warren
Wyoming

North Dakota

Adams
Billings
Bowman
Burleigh
Dickey
Dunn
Emmons

Golden Valley
Grant
Hettinger
LaMoure
Logan
McIntosh
McKenzie
Mercer
Morton
Oliver
Ransom
Richland
Sargent
Sioux
Slope
Stark

Pennsylvania

Cameron
Clearfield
Elk
McKean
Potter
Susquehanna
Tioga
Wayne

South Dakota

All counties except:
Bennett
Bon Homme
Charles Mix
Clay
Douglas
Gregory
Hutchinson
Jackson
Mellette
Todd
Tripp
Union
Yankton

Utah

Box Elder
Cache
Carbon
Daggett

Duchesne
Morgan
Rich
Summit
Uintah
Wasatch

Vermont

All counties

Washington

Ferry
Okanogan
Pend Oreille
Stevens

Wisconsin

All counties except:
Ashland
Bayfield
Burnett
Douglas
Florence
Forest
Iron
Langlade
Lincoln
Oneida
Price
Sawyer
Taylor
Vilas
Washburn

Wyoming

All counties except:
Goshen
Platte
Lincoln
Sublette
Teton

Climate Zone 6 Recommendation Table for Small to Medium Office Buildings

	Item	Component	Recommendation	How-To Tips	✓
Envelope	Roofs	Insulation entirely above deck	R-30.0 c.i.	EN2, 17, 19, 21–22	
		Attic and other	R-49.0	EN3, 17, 19, 20–21	
		Metal building	R-25.0 + R-11.0 Ls	EN4, 17, 19, 21	
		SRI	No recommendation	None	
	Walls	Mass (HC > 7 Btu/ft^2)	R-19.0 c.i.	EN5, 17, 19, 21	
		Steel framed	R-13.0 + R-18.8 c.i.	EN6, 17, 19, 21	
		Wood framed and other	R-13.0 + R-12.5 c.i.	EN7, 17, 19, 21	
		Metal building	R-0.0 + R-19.0 c.i.	EN8, 17, 19, 21	
		Below-grade walls	R-10.0 c.i.	EN9, 17, 19, 21–22	
		Mass	R-16.7 c.i.	EN10, 17, 19, 21	
	Floors	Steel joint	R-38.0	EN11, 17, 19, 21	
		Wood framed and other	R-38.0	EN11, 17, 19, 21	
	Slabs	Unheated	R-20.0 for 24 in.	EN12, 14, 17, 19, 21–22	
		Heated	R-20.0 for 48 in.	EN13–14, 17, 19, 21–22	
	Doors	Swinging	U-0.50	EN15, 17–18	
		Nonswinging	U-0.50	EN16–17	
	Vestibules	At building entrance	Yes	EN18	
	Continuous Air Barriers	Continuous air barrier	Entire building envelope	EN17	
	Vertical Fenestration	WWR	20% to 40%	EN25; DL6	
		Window orientation	Area of W and E windows each less than area of S windows (N in southern hemisphere)	EN30–31	
		Exterior sun control (S, E, and W only)	No recommendation	None	
		Thermal transmittance	Nonmetal framing windows = U-0.35 Metal framing windows = U-0.39	EN23–24	
		SHGC	Nonmetal framing windows = 0.35 Metal framing windows = 0.38	EN24, 32–33; DL12	
Daylighting / Lighting	Daylighting	Light-to-solar-gain ratio	Minimum VT/SHGC = 1.10	EN34–42	
		Vertical fenestration EA	0.12	DL7–11	
	Interior Finishes	Interior surface average reflectance	Ceilings = 80% Wall surfaces = 70% Open office partitions = 50%	DL14–15; EL4	
		Open office partitions parallel to window walls	Total partition height = 36 in. maximum - or - Partition above desk height = min 50% translucent	DL4, 10; EL2, 4	
	Interior Lighting	LPD	0.75 W/ft^2	DL1–5; EL2–3, 5, 12–19	
		24-hour lighting LPD	0.075 W/ft^2	EL20	
		Light source lamp efficacy (mean LPW)	T8 and T5 lamps > 2 ft = 92 T8 and T5 lamps ≤ 2 ft = 85 All other > 50	EL6–8	
		Ballasts	4 ft T8 lamp nondimming applications – NEMA Premium instant start 4 ft T8 lamp dimming applications = NEMA Premium program start Fluorescent and HID sources = electronic	EL8	
		Controls for daylight harvesting in open offices—locate on N and S sides of bldg	Dim all general fluorescent lights within primary and secondary daylight zones of open offices	DL16–20; EL1, 11	
		Automatic controls	Auto ON to 50% = private offices, conference and meeting rooms, lounge and break rooms, copy rooms, storage rooms Auto ON occupancy sensors = restrooms, electrical/mechanical rooms, open and private office task lighting Time switch control = all other spaces	DL13; EL1, 5, 9–10	
	Exterior Lighting	Façade and landscape lighting	LPD = 0.075 W/ft^2 in LZ3 and LZ4, 0.05 W/ft^2 in LZ2 Controls = auto OFF between 12 am and 6 am	EL23	
		Parking Lots and drives	LPD = 0.1 W/ft^2 in LZ3 and LZ4, 0.06 W/ft^2 in LZ2 Controls = auto reduce to 25% (12 am to 6 am)	EL21, 24–25	
		Walkways, plazas, and special feature areas	LPD = 0.16 W/ft^2 LZ3 and LZ4, 0.14 W/ft^2 in LZ2 Controls = auto reduce to 25% (12 am to 6 am)	EL22, 24–25	
		All other exterior lighting	LPD = follow Standard 90.1-2010 Controls = auto reduce to 25% (12 am to 6 am)	EL24–25	
Plug Loads	Equipment Choices	Laptop computers	Minimum 2/3 of total computers	PL1–2	
		ENERGY STAR equipment	For all computers, equipment, and appliances	PL1–2	
		Equipment power density	For all computers, equipment, and appliances	PL1–2, 4–6	
	Controls	Computer power control	Network control with power saving modes and control OFF during unoccupied hours	PL3	
		Occupancy sensors	Desk plug strip occupancy sensors	PL3	
		Timer switches	Water coolers and coffee makers control OFF during unoccupied hours	PL3	
		Vending machine control	Yes	PL3	

*Note: Where the table says "No recommendation," the user must meet the more stringent of either the applicable version of ASHRAE/IES Standard 90.1 or the local code requirements.

Climate Zone 6 Recommendation Table for Small to Medium Office Buildings *(Continued)*

	Item	Component	Recommendation	How-To Tips	✓
SWH	SWH	Gas water heater efficiency	Condensing water heaters = 90% efficiency	WH1–5	
		Electric storage EF (≤12 kW, ≥20 gal)	EF > 0.99 – 0.0012 × volume (see Table 5-7)	WH1–5	
		Point-of-use heater selection	0.81 EF or 81% E_t	WH1–5	
		Electric heat pump water heater efficiency	COP 3.0 (interior heat source)	WH1–5	
		Pipe insulation (d < 1½ in. / d ≥ 1½ in.)	1 in. / 1 ½ in.	WH6	
HVAC	Packaged Single-Zone Air-Source Heat Pumps	Cooling and heating efficiency	See Table 5-8 for efficiency	HV3	
		ESP	0.7 in. w.c.	HV3	
		DOAS air-source heat pump	Yes	HV10	
		DOAS heating and cooling efficiency	See Table 5-10 for efficiency	HV10–11	
		DOAS energy recovery	Yes, see Table 5-11 for effectiveness	HV12	
		DOAS fan—ESP	1.5 in. w.c. maximum	HV20	
	WSHPs with DOAS	Unit size	5 tons or less	None	
		Single-stage cooling and heating efficiency	Cooling = 16.4 EER Heating = 5.2 COP	HV4	
		Two-stage cooling and heating efficiency	Cooling part load = 17.6 EER Cooling full load = 15.0 EER Heating part load = 5.7 COP Heating full load = 5.0 COP	HV4	
		WSHP fan—ESP	0.5 in. w.c.	HV4, 20	
		Condensing boiler efficiency	90%	HV14, 30	
		DOAS water-to-water heat pump	See Table 5-10 for efficiency	HV10–11	
		DOAS variable airflow with DCV	Yes	HV17–18	
		DOAS energy recovery	Yes, see Table 5-11 for effectiveness	HV12	
		DOAS fan and motor	65% mechanical efficiency Motor efficiency per Standard 90.1-2010, Table 10.8B VSD efficiency = 95%	HV23	
	VAV DX with Gas-Fired Hydronic Heating	DX efficiency	See Table 5-9 for efficiency	HV6	
		Low-temperature air supply and SAT reset	50°F to 61°F	HV6, 11, 25, 31	
		Perimeter convector heat source	Hot water	None	
		Condensing boiler efficiency	90%	HV14, 30	
		Economizer	≥54,000 Btu/h, differential dry-bulb control	HV6, 16	
		Energy recovery	Yes, see Table 5-11 for effectiveness	HV12	
		Indirect evaporative cooling	No recommendation	None	
		Demand control and ventilation reset	Yes	HV17–18	
		ESP	2.0 in. w.c.	HV7, 20	
	VAV CHW (same as VAV DX except...)	Air-cooled chiller efficiency	10 EER	HV14, 29	
		Air-cooled chiller IPLV	12.5 IPLV < 150 tons, 12.75 IPLV ≥ 150 tons	HV14	
		Variable-speed pumping	Yes	HV29	
		Maximum fan power	0.72 W/cfm	HV7, 23, 31	
	Fan-Coils with DOAS	Air-cooled chiller efficiency	10 EER	HV14, 29	
		Air-cooled chiller IPLV	12.5 IPLV < 150 tons 12.75 IPLV ≥ 150 tons	HV14	
		Condensing boiler efficiency	90%	HV14, 30	
		Variable-speed pumping	Yes	HV29–30	
		VAV fan-coil units	Yes	HV8	
		Fan-coil unit fan power	0.30 W/cfm	HV8	
		DOAS chilled-water and hot-water coils served by same plant as fan-coils	Yes	HV10	
		DOAS variable airflow with DCV	Yes	HV10–11, 17	
		DOAS energy recovery	Yes, see Table 5-11 for effectiveness	HV12	
		DOAS fan and motor	65% mechanical efficiency, motor efficiency Standard 90.1-2010, Table 10.8B	HV23	
	Radiant Systems with DOAS	Air-cooled chiller full-load efficiency	10 EER	HV14	
		Air-cooled chiller IPLV	12.5 IPLV < 150 tons, 12.75 IPLV ≥ 150 tons	HV14	
		Condensing boiler efficiency	90%	HV14, 30	
		DOAS heating and cooling efficiency	See Table 5-10 for efficiency	HV10–11	
		DOAS variable airflow with DCV	Yes	HV10–11, 17–18	
		DOAS energy recovery	Yes, see Table 5-11 for effectiveness	HV12	
		DOAS fan—ESP	1.5 in. w.c.	HV20	
	Ducts and Dampers	OA damper	Motorized damper	HV16	
		Friction rate	0.08 in./100 ft	HV20	
		Sealing	Seal Class B	HV22	
		Location	Interior only	HV20	
		Insulation level	R-6.0	HV21	

*Note: Where the table says "No recommendation," the user must meet the more stringent of either the applicable version of ASHRAE/IES Standard 90.1 or the local code requirements.

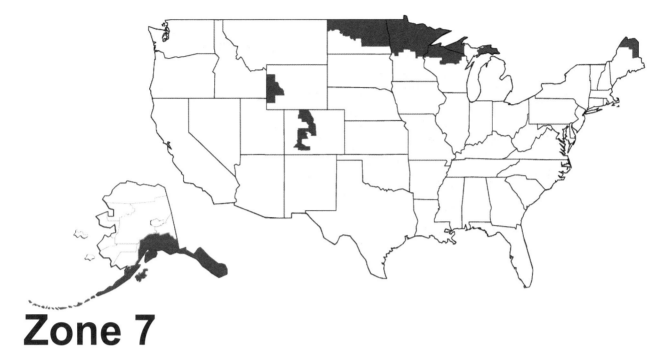

Zone 7

Alaska

Aleutians East
Aleutians West (CA)
Anchorage
Angoon (CA)
Bristol Bay
Denali
Haines
Juneau
Kenai Peninsula
Ketchikan (CA)
Ketchikan Gateway
Kodiak Island
Lake and Peninsula
Matanuska-Susitna
Prince of Wales-Outer
Sitka
Skagway-Hoonah-
Valdez-Cordova (CA)
Wrangell-Petersburg (CA)
Yakutat

Colorado

Clear Creek
Grand
Gunnison
Hinsdale
Jackson
Lake
Mineral
Park
Pitkin
Rio Grande
Routt
San Juan
Summit

Maine

Aroostook

Michigan

Baraga
Chippewa
Gogebic
Houghton
Iron

Keweenaw
Luce
Mackinac
Ontonagon
Schoolcraft

Minnesota

Aitkin
Becker
Beltrami
Carlton
Cass
Clay
Clearwater
Cook
Crow Wing
Grant
Hubbard
Itasca
Kanabec
Kittson
Koochiching
Lake
Lake of the Woods
Mahnomen
Marshall
Mille Lacs
Norman
Otter Tail
Pennington
Pine
Polk
Red Lake
Roseau
St. Louis
Wadena
Wilkin

North Dakota

Barnes
Benson
Bottineau
Burke
Cass
Cavalier
Divide
Eddy

Foster
Grand Forks
Griggs
Kidder
McHenry
McLean
Mountrail
Nelson
Pembina
Pierce
Ramsey
Renville
Rolette
Sheridan
Steele
Stutsman
Towner
Traill
Walsh
Ward
Wells
Williams

Wisconsin

Ashland
Bayfield
Burnett
Douglas
Florence
Forest
Iron
Langlade
Lincoln
Oneida
Price
Sawyer
Taylor
Vilas
Washburn

Wyoming

Lincoln
Sublette
Teton

Climate Zone 7 Recommendation Table for Small to Medium Office Buildings

	Item	Component	Recommendation	How-To Tips	✓
Envelope	Roofs	Insulation entirely above deck	R-35.0 c.i.	EN2, 17, 19, 21–22	
		Attic and other	R-60.0	EN3, 17, 19, 20–21	
		Metal building	R-30.0 + R-11.0 Ls	EN4, 17, 19, 21	
		SRI	No recommendation	None	
	Walls	Mass (HC > 7 Btu/ft^2)	R-19.0 c.i.	EN5, 17, 19, 21	
		Steel framed	R-13.0 + R-18.8 c.i.	EN6, 17, 19, 21	
		Wood framed and other	R-13.0 + R-15.0 c.i.	EN7, 17,19, 21	
		Metal building	R-0.0 + R-22.1 c.i.	EN8, 17, 19, 21	
		Below-grade walls	R-15.0 c.i.	EN9, 17, 19, 21–22	
	Floors	Mass	R-20.9 c.i.	EN10, 17, 19, 21	
		Steel joint	R-49.0	EN11, 17, 19, 21	
		Wood framed and other	R-49.0	EN11, 17, 19, 21	
	Slabs	Unheated	R-20.0 for 24 in.	EN12, 14, 17, 19, 21–22	
		Heated	R-25.0 for 48 in.	EN13,14, 17, 19, 21–22	
	Doors	Swinging	U-0.50	EN15, 17–18	
		Nonswinging	U-0.50	EN16–17	
	Vestibules	At building entrance	Yes	EN18	
	Continuous Air Barriers	Continuous air barrier	Entire building envelope	EN17	
	Vertical Fenestration	WWR	20% to 40%	EN25; DL6	
		Window orientation	Area of W and E windows each less than area of S windows (N in southern hemisphere)	EN30–31	
		Exterior sun control (S, E, and W only)	No recommendation	None	
		Thermal transmittance	Nonmetal framing windows = U-0.33 Metal framing windows = U-0.34	EN23–24	
		SHGC	Nonmetal framing windows = 0.40 Metal framing windows = 0.40	EN24, 32–33; DL12	
Daylighting / Lighting	Daylighting	Light-to-solar-gain ratio	Minimum VT/SHGC = 1.10	EN34–42	
		Vertical fenestration EA	0.12	DL7–11	
	Interior Finishes	Interior surface average reflectance	Ceilings = 80% Wall surfaces = 70% Open office partitions = 50%	DL14–15; EL4	
		Open office partitions parallel to window walls	Total partition height = 36 in. maximum - or - Partition above desk height = min 50% translucent	DL4, 10; EL2, 4	
	Interior Lighting	LPD	0.75 W/ft^2	DL1–5; EL2–3, 5, 12–19	
		24-hour lighting LPD	0.075 W/ft^2	EL20	
		Light source lamp efficacy (mean LPW)	T8 and T5 lamps > 2 ft = 92 T8 and T5 lamps ≤ 2 ft = 85 All other > 50	EL6–8	
		Ballasts	4 ft T8 lamp nondimming applications = NEMA Premium instant start 4 ft T8 lamp dimming applications = NEMA Premium program start Fluorescent and HID sources = electronic	EL8	
		Controls for daylight harvesting in open offices—locate on N and S sides of bldg	Dim all general fluorescent lights within primary and secondary daylight zones of open offices	DL16–20; EL1, 11	
		Automatic controls	Auto ON to 50% = private offices, conference and meeting rooms, lounge and break rooms, copy rooms, storage rooms Auto ON occupancy sensors = restrooms, electrical/ mechanical rooms, open and private office task lighting Time switch control = all other spaces	DL13; EL1, 5, 9–10	
	Exterior Lighting	Façade and landscape lighting	LPD = 0.075 W/ft^2 in LZ3 and LZ4, 0.05 W/ft^2 in LZ2 Controls = auto OFF between 12 am and 6 am	EL23	
		Parking lots and drives	LPD = 0.1 W/ft^2 in LZ3 and LZ4, 0.06 W/ft^2 in LZ2 Controls = auto reduce to 25% (12 am to 6 am)	EL21, 24–25	
		Walkways, plazas, and special feature areas	LPD = 0.16 W/ft^2 LZ3 and LZ4, 0.14 W/ft^2 in LZ2 Controls = auto reduce to 25% (12 am to 6 am)	EL22, 24–25	
		All other exterior lighting	LPD = follow Standard 90.1-2010 Controls = auto reduce to 25% (12 am to 6 am)	EL24–25	
Plug Loads	Equipment Choices	Laptop computers	Minimum 2/3 of total computers	PL1–2	
		ENERGY STAR equipment	For all computers, equipment, and appliances	PL1–2	
		Equipment power density	For all computers, equipment, and appliances	PL1–2, 4–6	
	Controls	Computer power control	Network control with power saving modes and control OFF during unoccupied hours	PL3	
		Occupancy sensors	Desk plug strip occupancy sensors	PL3	
		Timer switches	Water coolers and coffee makers control OFF during unoccupied hours	PL3	
		Vending machine control	Yes	PL3	

*Note: Where the table says "No recommendation," the user must meet the more stringent of either the applicable version of ASHRAE/IES Standard 90.1 or the local code requirements.

Climate Zone 7 Recommendation Table for Small to Medium Office Buildings *(Continued)*

	Item	Component	Recommendation	How-To Tips	✓
SWH	SWH	Gas water heater efficiency	Condensing water heaters = 90% efficiency	WH1–5	
		Electric storage EF (≤12 kW, ≥20 gal)	EF > 0.99 – 0.0012 × volume (see Table 5-7)	WH1–5	
		Point-of-use heater selection	0.81 EF or 81% E_t	WH1–5	
		Electric heat pump water heater efficiency	COP 3.0 (interior heat source)	WH1–5	
		Pipe insulation (d < 1½ in. / d ≥ 1½ in.)	1 in. / 1 ½ in.	WH6	
HVAC	Packaged Single-Zone Air-Source Heat Pumps	Cooling and heating efficiency	See Table 5-8 for efficiency	HV3	
		ESP	0.7 in. w.c.	HV3	
		DOAS air-source heat pump	Yes	HV10	
		DOAS heating and cooling efficiency	See Table 5-10 for efficiency	HV10–11	
		DOAS energy recovery	Yes, see Table 5-11 for effectiveness	HV12	
		DOAS fan—ESP	1.5 in. w.c. maximum	HV20	
		Unit size	5 tons or less	None	
	WSHPs with DOAS	Single-stage cooling and heating efficiency	Cooling = 16.4 EER Heating = 5.2 COP	HV4	
		Two-stage cooling and heating efficiency	Cooling part load = 17.6 EER Cooling full load = 15.0 EER Heating part load = 5.7 COP Heating full load = 5.0 COP	HV4	
		WSHP fan—ESP	0.5 in. w.c.	HV4, 20	
		Condensing boiler efficiency	90%	HV14, 30	
		DOAS water-to-water heat pump	See Table 5-10 for efficiency	HV10–11	
		DOAS variable airflow with DCV	Yes	HV17–18	
		DOAS energy recovery	Yes, see Table 5-11 for effectiveness	HV12	
		DOAS fan and motor	65% mechanical efficiency Motor efficiency per Standard 90.1-2010, Table 10.8B VSD efficiency = 95%	HV23	
	VAV DX with Gas-Fired Hydronic Heating	DX efficiency	See Table 5-9 for efficiency	HV6	
		Low-temperature air supply and SAT reset	50°F to 61°F	HV6, 11, 25, 31	
		Perimeter convector heat source	Hot water	None	
		Condensing boiler efficiency	90%	HV14, 30	
		Economizer	≥54,000 Btu/h, differential dry-bulb control	HV6, 16	
		Energy recovery	Yes, see Table 5-11 for effectiveness	HV12	
		Indirect evaporative cooling	No recommendation	None	
		Demand control and ventilation reset	Yes	HV17–18	
		ESP	2.0 in. w.c.	HV7, 20	
	VAV CHW (same as VAV DX except...)	Air-cooled chiller efficiency	10 EER	HV14, 29	
		Air-cooled chiller IPLV	12.5 IPLV < 150 tons, 12.75 IPLV ≥ 150 tons	HV14	
		Variable-speed pumping	Yes	HV29	
		Maximum fan power	0.72 W/cfm	HV7, 23, 31	
	Fan-Coils with DOAS	Air-cooled chiller efficiency	10 EER	HV14, 29	
		Air-cooled chiller IPLV	12.5 IPLV < 150 tons 12.75 IPLV ≥ 150 tons	HV14	
		Condensing boiler efficiency	90%	HV14, 30	
		Variable-speed pumping	Yes	HV29–30	
		VAV fan-coil units	Yes	HV8	
		Fan-coil unit fan power	0.30 W/cfm	HV8	
		DOAS chilled-water and hot-water coils served by same plant as fan-coils	Yes	HV10	
		DOAS variable airflow with DCV	Yes	HV10–11, 17	
		DOAS energy recovery	Yes, see Table 5-11 for effectiveness	HV12	
		DOAS fan and motor	65% mechanical efficiency, motor efficiency Standard 90.1-2010, Table 10.8B	HV23	
	Radiant Systems with DOAS	Air-cooled chiller full-load efficiency	10 EER	HV14	
		Air-cooled chiller IPLV	12.5 IPLV < 150 tons, 12.75 IPLV ≥ 150 tons	HV14	
		Condensing boiler efficiency	90%	HV14, 30	
		DOAS heating and cooling efficiency	See Table 5-10 for efficiency	HV10–11	
		DOAS variable airflow with DCV	Yes	HV10–11, 17–18	
		DOAS energy recovery	Yes, see Table 5-11 for effectiveness	HV12	
		DOAS fan—ESP	1.5 in. w.c.	HV20	
	Ducts and Dampers	OA damper	Motorized damper	HV16	
		Friction rate	0.08 in./100 ft	HV20	
		Sealing	Seal Class B	HV22	
		Location	Interior only	HV20	
		Insulation level	R-6.0	HV21	

*Note: Where the table says "No recommendation," the user must meet the more stringent of either the applicable version of ASHRAE/IES Standard 90.1 or the local code requirements.

Zone 8

Alaska

Bethel (CA)
Dillingham (CA)
Fairbanks North Star
Nome (CA)
North Slope
Northwest Arctic
Southeast Fairbanks (CA)
Wade Hampton (CA)
Yukon-Koyukuk (CA)

Climate Zone 8 Recommendation Table for Small to Medium Office Buildings

	Item	Component	Recommendation	How-To Tips	✓
Envelope	Roofs	Insulation entirely above deck	R-35.0 c.i.	EN2, 17, 19, 21–22	
		Attic and other	R-60.0	EN3, 17, 19, 20–21	
		Metal building	R-25.0 + R-11.0 + R-11.0 Ls	EN4, 17, 19, 21	
		SRI	No recommendation	None	
	Walls	Mass (HC > 7 Btu/ft^2)	R-19.0 c.i.	EN5, 17, 19, 21	
		Steel framed	R-13.0 + R-18.8 c.i.	EN6, 17, 19, 21	
		Wood framed and other	R-13.0 + R-18.8 c.i.	EN7, 17, 19, 21	
		Metal building	R-0.0 + R-25.0 c.i.	EN8, 17, 19, 21	
		Below-grade walls	R-15.0 c.i.	EN9, 17, 19, 21–22	
	Floors	Mass	R-23.0 c.i.	EN10, 17, 19, 21	
		Steel joint	R-60.0	EN11, 17, 19, 21	
		Wood framed and other	R-60.0	EN11, 17, 19, 21	
	Slabs	Unheated	R-20.0 for 48 in.	EN12, 14, 17, 19, 21–22	
		Heated	R-20.0 full slab	EN13–14, 17, 19, 21–22	
	Doors	Swinging	U-0.50	EN15, 17–18	
		Nonswinging	U-0.50	EN16–17	
	Vestibules	At building entrance	Yes	EN18	
	Continuous Air Barriers	Continuous air barrier	Entire building envelope	EN17	
	Vertical Fenestration	WWR	20% to 40%	EN25; DL6	
		Window orientation	Area of W and E windows each less than area of S windows (N in southern hemisphere)	EN30–31	
		Exterior sun control (S, E, and W only)	No recommendation	None	
		Thermal transmittance	Nonmetal framing windows = U-0.25 Metal framing windows = U-0.34	EN23–24	
		SHGC	Nonmetal framing windows = 0.40 Metal framing windows = 0.40	EN24, 32–33; DL12	
Daylighting / Lighting	Daylighting	Light-to-solar-gain ratio	Minimum VT/SHGC = 1.10	EN34–42	
		Vertical fenestration EA	0.12	DL7–11	
	Interior Finishes	Interior surface average reflectance	Ceilings = 80% Wall surfaces = 70% Open office partitions = 50%	DL14–15; EL4	
		Open office partitions parallel to window walls	Total partition height = 36 in. maximum - or - Partition above desk height = min 50% translucent	DL4, 10; EL2, 4	
	Interior Lighting	LPD	0.75 W/ft^2	DL1–5; EL2–3, 5, 12–19	
		24-hour lighting LPD	0.075 W/ft^2	EL20	
		Light source lamp efficacy (mean LPW)	T8 and T5 lamps > 2 ft = 92 T8 and T5 lamps ≤ 2 ft = 85 All other > 50	EL6–8	
		Ballasts	4 ft T8 lamp nondimming applications = NEMA Premium instant start 4 ft T8 lamp dimming applications = NEMA Premium program start Fluorescent and HID sources = electronic	EL8	
		Controls for daylight harvesting in open offices— locate on N and S sides of bldg	Dim all general fluorescent lights within primary and secondary daylight zones of open offices	DL16–20; EL1, 11	
		Automatic controls	Auto ON to 50% = private offices, conference and meeting rooms, lounge and break rooms, copy rooms, storage rooms Auto ON occupancy sensors = restrooms, electrical/ mechanical rooms, open and private office task lighting Time switch control = all other spaces	DL13; EL1, 5, 9–10	
	Exterior Lighting	Façade and landscape lighting	LPD = 0.075 W/ft^2 in LZ3 and LZ4, 0.05 W/ft^2 in LZ2 Controls = auto OFF between 12 am and 6 am	EL23	
		Parking lots and drives	LPD = 0.1 W/ft^2 in LZ3 and LZ4, 0.06 W/ft^2 in LZ2 Controls = auto reduce to 25% (12 am to 6 am)	EL21, 24–25	
		Walkways, plazas, and special feature areas	LPD = 0.16 W/ft^2 LZ3 and LZ4, 0.14 W/ft^2 in LZ2 Controls = auto reduce to 25% (12 am to 6 am)	EL22, 24–25	
		All other exterior lighting	LPD = follow 90.1-2010 Controls = auto reduce to 25% (12am to 6 am)	EL24–25	
Plug Loads	Equipment Choices	Laptop computers	Minimum 2/3 of total computers	PL1–2	
		ENERGY STAR equipment	For all computers, equipment, and appliances	PL1–2	
		Equipment power density	For all computers, equipment, and appliances	PL1–2, 4–6	
	Controls	Computer power control	Network control with power saving modes and control OFF during unoccupied hours	PL3	
		Occupancy sensors	Desk plug strip occupancy sensors	PL3	
		Timer switches	Water coolers and coffee makers control OFF during unoccupied hours	PL3	
		Vending machine control	Yes	PL3	

*Note: Where the table says "No recommendation," the user must meet the more stringent of either the applicable version of ASHRAE/IES Standard 90.1 or the local code requirements.

Climate Zone 8 Recommendation Table for Small to Medium Office Buildings *(Continued)*

	Item	Component	Recommendation	How-To Tips	✓
SWH	SWH	Gas water heater efficiency	Condensing water heaters = 90% efficiency	WH1–5	
		Electric storage EF (≤12 kW, ≥20 gal)	EF > 0.99 – 0.0012 × volume (see Table 5-7)	WH1–5	
		Point-of-use heater selection	0.81 EF or 81% E_t	WH1–5	
		Electric heat pump water heater efficiency	COP 3.0 (interior heat source)	WH1–5	
		Pipe insulation (d < 1½ in. / d ≥ 1½ in.)	1 in. / 1 ½ in.	WH6	
HVAC	Packaged Single-Zone Air-Source Heat Pumps	Cooling and heating efficiency	See Table 5-8 for efficiency	HV3	
		ESP	0.7 in. w.c.	HV3	
		DOAS air-source heat pump	Yes	HV10	
		DOAS heating and cooling efficiency	See Table 5-10 for efficiency	HV10–11	
		DOAS energy recovery	Yes, see Table 5-11 for effectiveness	HV12	
		DOAS fan—ESP	1.5 in. w.c. maximum	HV20	
	WSHPs with DOAS	Unit size	5 tons or less	None	
		Single-stage cooling and heating efficiency	Cooling = 16.4 EER Heating = 5.2 COP	HV4	
		Two-stage cooling and heating efficiency	Cooling part load = 17.6 EER Cooling full load = 15.0 EER Heating part load = 5.7 COP Heating full load = 5.0 COP	HV4	
		WSHP fan—ESP	0.5 in. w.c.	HV4, 20	
		Condensing boiler efficiency	90%	HV14, 30	
		DOAS water-to-water heat pump	See Table 5-10 for efficiency	HV10–11	
		DOAS variable airflow with DCV	Yes	HV17–18	
		DOAS energy recovery	Yes, see Table 5-11 for effectiveness	HV12	
		DOAS fan and motor	65% mechanical efficiency Motor efficiency per Standard 90.1-2010, Table 10.8B VSD efficiency = 95%	HV23	
	VAV DX with Gas-Fired Hydronic Heating	DX efficiency	See Table 5-9 for efficiency	HV6	
		Low-temperature air supply and SAT reset	50°F to 61°F	HV6, 11, 25, 31	
		Perimeter convector heat source	Hot water	None	
		Condensing boiler efficiency	90%	HV14, 30	
		Economizer	≥54,000 Btu/h, differential dry-bulb control	HV6, 16	
		Energy recovery	Yes, see Table 5-11 for effectiveness	HV12	
		Indirect evaporative cooling	No recommendation	None	
		Demand control and ventilation reset	Yes	HV17–18	
		ESP	2.0 in. w.c.	HV7, 20	
	VAV CHW (same as VAV DX except...)	Air-cooled chiller efficiency	10 EER	HV14, 29	
		Air-cooled chiller IPLV	12.5 IPLV < 150 tons, 12.75 IPLV ≥ 150 tons	HV14	
		Variable-speed pumping	Yes	HV29	
		Maximum fan power	0.72 W/cfm	HV7, 23, 31	
	Fan-Coils with DOAS	Air-cooled chiller efficiency	10 EER	HV14, 29	
		Air-cooled chiller IPLV	12.5 IPLV < 150 tons 12.75 IPLV ≥ 150 tons	HV14	
		Condensing boiler efficiency	90%	HV14, 30	
		Variable-speed pumping	Yes	HV29–30	
		VAV fan-coil units	Yes	HV8	
		Fan-coil unit fan power	0.30 W/cfm	HV8	
		DOAS chilled-water and hot-water coils served by same plant as fan-coils	Yes	HV10	
		DOAS variable airflow with DCV	Yes	HV10–11, 17	
		DOAS energy recovery	Yes, see Table 5-11 for effectiveness	HV12	
		DOAS fan and motor	65% mechanical efficiency, motor efficiency Standard 90.1-2010, Table 10.8B	HV23	
	Radiant Systems with DOAS	Air-cooled chiller full-load efficiency	10 EER	HV14	
		Air-cooled chiller IPLV	12.5 IPLV < 150 tons, 12.75 IPLV ≥ 150 tons	HV14	
		Condensing boiler efficiency	90%	HV14, 30	
		DOAS heating and cooling efficiency	See Table 5-10 for efficiency	HV10–11	
		DOAS variable airflow with DCV	Yes	HV10–11, 17–18	
		DOAS energy recovery	Yes, see Table 5-11 for effectiveness	HV12	
		DOAS fan—ESP	1.5 in. w.c.	HV20	
	Ducts and Dampers	OA damper	Motorized damper	HV16	
		Friction rate	0.08 in./100 ft	HV20	
		Sealing	Seal Class B	HV22	
		Location	Interior only	HV20	
		Insulation level	R-6.0	HV21	

*Note: Where the table says "No recommendation," the user must meet the more stringent of either the applicable version of ASHRAE/IES Standard 90.1 or the local code requirements.

REFERENCES

ASHRAE. 2004. ANSI/ASHRAE/IESNA Standard 90.1-2004, *Energy Standard for Buildings Except Low-Rise Residential Buildings*. Atlanta: American Society of Heating, Refrigerating and Air-Conditioning.

ASHRAE. 2010a. ANSI/ASHRAE/IES Standard 90.1-2010, *Energy Standard for Buildings Except Low-Rise Residential Buildings*. Atlanta: American Society of Heating, Refrigerating and Air-Conditioning.

ASHRAE. 2010ba. ANSI/ASHRAE Standard 62.1-2010, *Ventilation for Acceptable Indoor Air Quality*. Atlanta: American Society of Heating, Refrigerating and Air-Conditioning.

Briggs, R.S., R.G. Lucas, and Z.T. Taylor. 2003. Climate classification for building energy codes and standards: Part 1—Development process. *ASHRAE Transactions* 109(1):109–121.

Murphy, J. 2008. Energy-saving strategies for rooftop VAV systems. *HPAC Engineering*, May, pp. 28–35.

Stanke, D. 2006. Standard 62.1-2004 system operation: Dynamic reset options. *ASHRAE Journal* 48(12):18–32.

Stanke, D. 2010. Dynamic reset for multiple-zone systems. *ASHRAE Journal* 52(3):22–35.

Taylor, S.T. 2006. CO_2-based DCV using 62.1-2004. *ASHRAE Journal* 48(5):71–77.

How to Implement Recommendations

5

Recommendations for energy-saving measures for each climate zone are contained in the individual tables in Chapter 4, "Strategies and Recommendations by Climate Zone." The following how-to tips are intended to provide guidance on good practices for implementing the recommendations as well as cautions to avoid known problems in energy-efficient construction.

ENVELOPE

OPAQUE ENVELOPE COMPONENTS

Good Design Practice

EN1   *Cool Roofs* **(Climate Zones: ❶ ❷ ❸)**

For a roof to be considered a cool roof, a Solar Reflectance Index (SRI) of 78 or higher is recommended. A high reflectance keeps much of the sun's energy from being absorbed while a high thermal emissivity surface radiates away any solar energy that is absorbed, allowing the roof to cool more rapidly. Cool roofs are typically white and have a smooth surface. Commercial roof products that qualify as cool roofs fall into three categories: single ply, liquid applied, and metal panels. Examples are presented in Table 5-1.

Table 5-1   Examples of Cool Roofs

Category	Product	Reflectance	Emissivity	SRI
Single ply	White polyvinyl chloride (PVC)	0.86	0.86	107
	White chlorinated polyethylene (CPE)	0.86	0.88	108
	White chlorosulfonated polyethylene (CPSE)	0.85	0.87	106
	White thermoplastic polyolefin (TSO)	0.77	0.87	95
Liquid applied	White elastomeric, polyurethane, acrylic coating	0.71	0.86	86
	White paint (on metal or concrete)	0.71	0.85	86
Metal panels	Factory-coated white finish	0.90	0.87	113

The solar reflectance and thermal emissivity property values represent initial conditions as determined by a laboratory accredited by the Cool Roof Rating Council (CRRC). An SRI can be determined by the following equation:

$$SRI = 123.97 - 141.35(\chi) + 9.655(\chi^2)$$

where

$$\chi = \frac{20.797 \times \alpha - 0.603 \times \varepsilon}{9.5205 \times \varepsilon + 12.0}$$

and

α = solar absorptance = 1 – solar reflectance

ε = thermal emissivity

These equations were derived from ASTM E1980 (ASTM 2011) assuming a medium wind speed. Note that cool roofs are not a substitute for the appropriate amount of insulation.

EN2 *Roofs—Insulation Entirely above Deck* (Climate Zones: all)

The insulation entirely above deck should be continuous insulation (c.i.) rigid boards. Continuous insulation is important because no framing members are present that would introduce thermal bridges or short circuits to bypass the insulation. When two layers of c.i. are used in this construction, the board edges should be staggered to reduce the potential for convection losses or thermal bridging. If an inverted or protected membrane roof system is used, at least one layer of insulation is placed above the membrane and a maximum of one layer is placed beneath the membrane.

EN3 *Roofs—Attics and Other Roofs* (Climate Zones: all)

Attics and other roofs include roofs with insulation entirely below (inside of) the roof structure (i.e., attics and cathedral ceilings) and roofs with insulation both above and below the roof structure. Ventilated attic spaces need to have the insulation installed at the ceiling line. Unventilated attic spaces may have the insulation installed at the roof line. When suspended ceilings with removable ceiling tiles are used, the insulation performance is best when installed at the roof line. For buildings with attic spaces, ventilation should be provided equal to 1 ft² of open area per 100 ft² of attic space. This will provide adequate ventilation as long as the openings are split between the bottom and top of the attic space. (See Figure 5-1.)

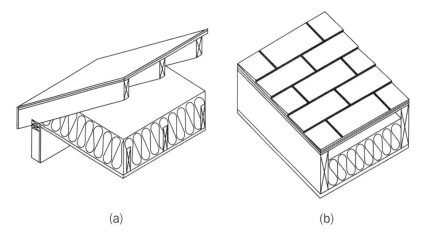

(a)   (b)

Figure 5-1 (EN3) Attics and Other Roofs—(a) Ventilated Attic and (b) Cathedral Ceiling

EN4   Roofs—Metal Buildings (Climate Zones: all)

Metal buildings pose particular challenges in the pursuit of designing and constructing advanced buildings. The metal skin and purlin/girt connection, even with compressed fiberglass between the two, is highly conductive, which limits the effectiveness of the insulation. A purlin is a horizontal structural member that supports the roof covering. In metal building construction, this is typically a z-shaped cold-formed steel member; a steel bar or open web joists can be used for longer spans.

The thermal performance of metal building roofs with fiberglass batts is improved by treating the thermal bridging associated with fasteners. Use of foam blocks is a proven technique to reduce thermal bridging. Thermal blocks, with minimum dimensions of 1 × 3 in., should be R-5 rigid insulation installed parallel to the purlins. (See Figure 5-2.)

Thermal blocks can be used successfully with standing seam roofs that use concealed clips of varying heights to accommodate the block. However, a thermal block cannot be used with a through-fastened roof that is screwed directly to the purlins because it diminishes the structural load carrying capacity by "softening" the connection and restraint provided to the purlin by the roof.

In climate zones 1 through 3, the recommended construction is a filled cavity that has the first layer of insulation, R-10, perpendicular to and over the top of the purlins and the second layer of insulation, R-19, parallel to and between the purlins (see Figure 5-2a).

In climate zones 4 through 7, the recommended construction is a liner system that has the first layer of insulation parallel to and between the purlins and the second layer of insulation perpendicular to and over the top of the purlins (see Figure 5-2b).

In climate zone 8, the recommended construction is a liner system with the first and second layers of insulation parallel to and between the purlins and the third layer of insulation perpendicular to and over the top of the purlins (see Figure 5-2c).

Rigid c.i. can be added to provide additional insulation if required to meet the U-factors listed in Appendix A. In any case, rigid c.i. or other high-performance insulation systems may be used provided the total roof assembly has a U-factor that is less than or equal to the appropriate climate zone construction listed in Appendix A.

EN5   Walls—Mass (Climate Zones: all)

Mass walls are defined as those with a heat capacity (HC) exceeding 7 Btu/ft^2·°F. Insulation may be placed on either the inside or the outside of the masonry wall. When insulation is placed on the exterior, rigid c.i. is recommended. When insulation is placed on the interior, a furring or framing system may be used, provided the total wall assembly has a U-factor that is less than or equal to the appropriate climate zone construction listed in Appendix A. (See Figure 5-3.)

The greatest advantages of mass walls can be obtained when insulation is placed on its exterior. In this case, the mass absorbs heat from the interior spaces that is later released in the

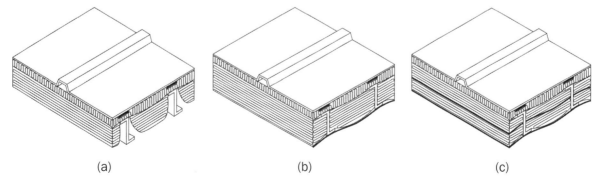

(a)   (b)   (c)

**Figure 5-2   (EN4) Prefabricated Metal Roofs Showing Thermal Blocking of Purlins—
(a) Filled Cavity; (b) Liner System, One Layer; and (c) Liner System, Two Layers**

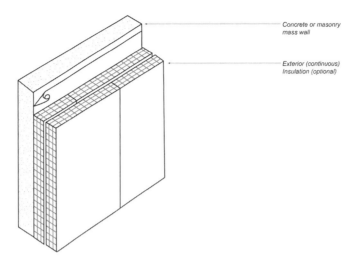

Figure 5-3 (EN5) Example Mass Wall Assembly

evenings when the buildings are not occupied. The thermal mass of a building (typically contained in the building envelope) absorbs heat during the day and reduces the magnitude of indoor air temperature swings, reduces peak cooling loads, and transfers some of the absorbed heat into the night hours. The cooling load can then be covered by passive cooling techniques (natural ventilation) when the outdoor conditions are more favorable. An unoccupied building can also be precooled during the night by natural or mechanical ventilation to reduce the cooling energy use. This same effect reduces heating load as well.

Thermal mass also has a positive effect on thermal comfort. High-mass buildings attenuate interior air and wall temperature variations and sustain a stable overall thermal environment. This increases thermal comfort, particularly during mild seasons (spring and fall), during large air temperature changes (high solar gain), and in areas with large day-night temperature swings.

Designers should keep in mind that the occupants will be the final determinants on the extent of the usability of any building system, including thermal mass. Changing the use of internal spaces and surfaces can drastically reduce the effectiveness of thermal storage. The final use of the space must be considered when doing heating and cooling load calculations and incorporating possible energy savings from thermal mass effects.

EN6   *Walls—Steel Framed* **(Climate Zones: all)**

Cold-formed steel framing members are thermal bridges to the cavity insulation. Adding exterior foam sheathing as c.i. is the preferred method to upgrade the wall thermal performance because it will increase the overall wall thermal performance and tends to minimize the impact of the thermal bridging.

Alternative combinations of cavity insulation and sheathing in thicker steel-framed walls can be used provided that the proposed total wall assembly has a U-factor that is less than or equal to the U-factor for the appropriate climate zone construction listed in Appendix A. Batt insulation installed in cold-formed steel-framed wall assemblies is to be ordered as "full width batt," and installation is normally by friction fit. Batt insulation should fill the entire cavity and not be cut short. (See Figure 5-4.)

EN7   *Walls—Wood Frame and Other* **(Climate Zones: all)**

Cavity insulation is used within the wood-frame wall, while rigid c.i. is placed on the exterior side of the framing (see Figure 5-5). Care must be taken to have a vapor retarder on the warm side of the wall and to use a vapor-retarder-faced batt insulation product to avoid insulation sagging away from the vapor retarder.

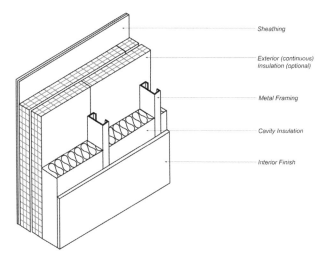

Figure 5-4   (EN6) Example Steel Frame Assembly

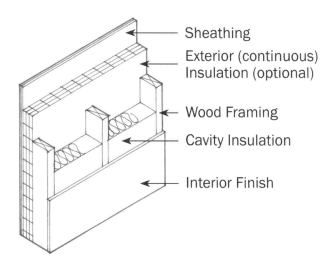

Figure 5-5   (EN7) Wood Frame and Other Walls

Alternative combinations of cavity insulations and sheathings in thicker walls can be used provided the total wall assembly has a U-factor that is less than or equal to the appropriate climate zone construction listed in Appendix A. Batt insulation should fill the entire cavity and not be cut short.

EN8   *Walls—Metal Building* (Climate Zones: all)

In climate zones where a single layer of fiberglass batt insulation is recommended, the insulation is installed continuously perpendicular to the exterior of the girts and is compressed as the metal panel is attached to the girts (see Figure 5-6). In climate zones where a layer of faced fiberglass batt insulation and a layer of rigid board insulation are recommended, the layer of fiberglass is installed continuously perpendicular to the exterior of the girts and is compressed as the rigid board insulation is installed continuously and perpendicular then attached to the girts from the exterior (on top of the fiberglass). The metal panels are then attached over the rigid board insulation using screws that penetrate the insulation assembly into the girts.

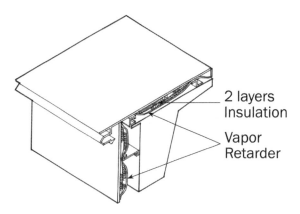

Figure 5-6   (EN8) Metal Building Walls

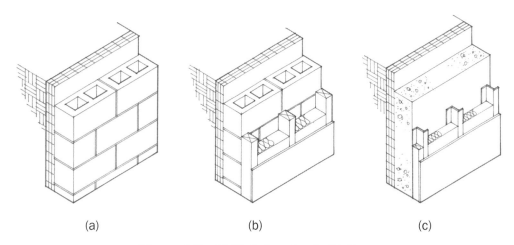

(a)   (b)   (c)

**Figure 5-7   (EN9) Below-Grade Walls—
(a) Exterior Insulation, (b) Interior Wood Framing, and (c) Interior Steel Framing**
The outer surface of the wall is in contact with the earth,
and the inside surface is adjacent to conditioned or semi-heated space.

In all climate zones, rigid c.i. is another option provided the total wall assembly has a U-factor that is less than or equal to the appropriate climate zone construction listed in Appendix A.

EN9   *Walls—Below-Grade* (Climate Zones: all)

Insulation, when recommended, may be placed on either the inside or the outside of a below-grade wall. If placed on the exterior of the wall, rigid c.i. is recommended. If placed on the interior of the wall, a furring or framing system is recommended, provided the total wall assembly has a C-factor that is less than or equal to the appropriate climate zone construction listed in Appendix A. (See Figure 5-7.)

EN10   *Floors—Mass* (Climate Zones: all)

Insulation should be continuous and either integral to or above the slab. This can be achieved by placing high-density extruded polystyrene above the slab with either plywood or a thin layer of concrete on top. Placing insulation below the deck is not recommended due to losses through any concrete support columns or through the slab perimeter. (See Figure 5-8.)

Exception: Buildings or zones within buildings that have durable floors for heavy machinery or equipment could place insulation below the deck.

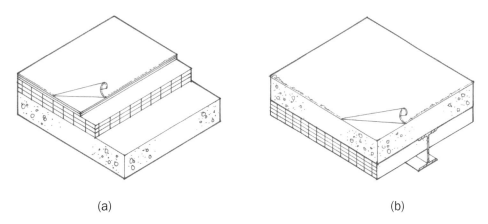

(a)   (b)

Figure 5-8   (EN10) Mass Floors—(a) Insulation Above Slab and (b) Insulation Below Slab
Any floor with a HC exceeding 7 $Btu/ft^2 \cdot °F$.

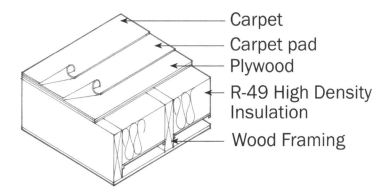

— Carpet
— Carpet pad
— Plywood
— R-49 High Density Insulation
— Wood Framing

Figure 5-9   (EN11) Wood-Frame Floors

When heated slabs are placed below grade, below-grade walls should meet the insulation recommendations for perimeter insulation according to the heated slab-on-grade construction.

EN11   Floors—Metal Joist or Wood Joist/Wood Frame (Climate Zones: all)

Insulation should be installed parallel to the framing members and in intimate contact with the flooring system supported by the framing member in order to avoid the potential thermal short-circuiting associated with open or exposed air spaces. Nonrigid insulation should be supported from below, no less frequently than at 24 in. on center. (See Figure 5-9.)

EN12   Slab-on-Grade Floors, Unheated (Climate Zones: all)

Rigid c.i. should be used around the perimeter of the slab and should reach the depth listed in the recommendation or to the bottom of the footing, whichever is less. In climate zone 8, c.i. should be placed beneath the slab as well. (See Figure 5-10.)

EN13   Slab-on-Grade Floors, Heated (Climate Zones: all)

Rigid c.i. should be used around the perimeter of the slab and should reach to the depth listed or to the frost line, whichever is deeper. Additionally, in climate zone 8, c.i. should be placed below the slab as well. (See Figure 5-11.)

Note: In areas where termites are a concern and rigid insulation is not recommended for use under the slab, a different heating system should be used.

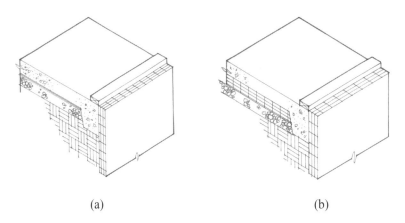

(a)   (b)

**Figure 5-10   (EN12) Slab-on-Grade Floors, Unheated—
(a) Perimeter Insulation and (b) Insulation Below the Slab**
No heating elements either within or below the slab.

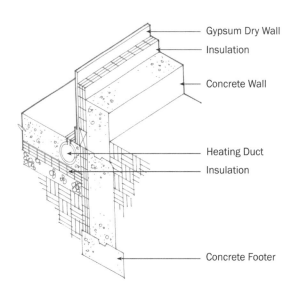

Figure 5-11   (EN13) Slab-on-Grade Floors, Heated
Heating elements either within (as shown) or below the slab.

EN14   ***Slab Edge Insulation* (Climate Zones: all)**

Use of slab edge insulation improves thermal performance, but problems can occur in regions that have termites.

EN15   ***Doors—Opaque, Swinging* (Climate Zones: all)**

A U-factor of 0.37 corresponds to an insulated double-panel metal door. A U-factor of 0.61 corresponds to a double-panel metal door. If at all possible, single swinging doors should be used. Double swinging doors are difficult to seal at the center of the doors unless there is a center post (see Figure 5-12). Double swinging doors without a center post should be minimized and limited to areas where width is important. Vestibules or revolving doors can be added to further improve energy efficiency.

EN16   ***Doors—Opaque, Roll-Up, or Sliding* (Climate Zones: all)**

Roll-up or sliding doors are recommended to have R-4.75 rigid insulation or meet the recommended U-factor. When meeting the recommended U-factor, the thermal bridging at the

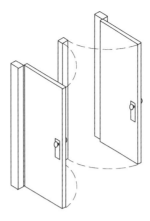

Figure 5-12 (EN15) Double Doors with a Center Post
Opaque doors with hinges on one side.

door and section edges is to be included in the analysis. Roll-up doors that have solar exposure should be painted with a reflective paint (or should be high emissivity) and should be shaded. Metal doors are a problem in that they typically have poor emissivity and collect heat, which is transmitted through even the best insulated door and causes cooling loads and thermal comfort issues.

If at all possible, use insulated panel doors over roll-up doors, as the insulation values can approach R-10 and provide a tighter seal to minimize infiltration.

EN17   *Air Infiltration Control* (**Climate Zones: all**)

The building envelope should be designed and constructed with a continuous air barrier system to control air leakage into or out of the conditioned space and should extend over all surfaces of the building envelope (at the lowest floor, exterior walls, and ceiling or roof). An air barrier system should also be provided for interior separations between conditioned space and space designed to maintain temperature or humidity levels that differ from those in the conditioned space by more than 50% of the difference between the conditioned space and design ambient conditions. If possible, a blower door should be used to depressurize the building to find leaks in the infiltration barrier. At a minimum, the air barrier system should have the following characteristics.

- It should be continuous, with all joints made airtight.
- Air barrier materials used in frame walls should have an air permeability not to exceed 0.004 cfm/ft^2 under a pressure differential of 0.3 in. w.c. (1.57 lb/ft^2) when tested in accordance with ASTM E 2178 (ASTM 2003).
- The system should be able to withstand positive and negative combined design wind, fan, and stack pressures on the envelope without damage or displacement and should transfer the load to the structure. It should not displace adjacent materials under full load.
- It should be durable or maintainable.
- The air barrier material of an envelope assembly should be joined in an airtight and flexible manner to the air barrier material of adjacent assemblies, allowing for the relative movement of these assemblies and components due to thermal and moisture variations, creep, and structural deflection.
- Connections should be made between the following:
  - Foundation and walls
  - Walls and windows or doors
  - Different wall systems
  - Wall and roof
  - Wall and roof over unconditioned space

- Walls, floors, and roof across construction, control, and expansion joints
- Walls, floors, and roof to utility, pipe, and duct penetrations
- All penetrations of the air barrier system and paths of air infiltration/exfiltration should be made airtight.

EN18 Vestibules (Climate Zones: ❸ ⦿ ❺ ❻ ❼ ❽)

Vestibules are recommended for building entrances routinely used by occupants, not for emergency exits, maintenance doors, loading docks, or any other specialty entrances. Occupant entrances that separate conditioned space from the exterior shall be protected with an enclosed vestibule, with all doors opening into and out of the vestibule equipped with self-closing devices. Vestibules shall be designed so that in passing through the vestibule it is not necessary for the interior and exterior doors to open at the same time. Interior and exterior doors shall have a minimum distance between them of not less than 7 ft when in the closed position. Vestibules shall be designed only as areas to traverse between the exterior and the interior. The exterior envelope of conditioned vestibules shall comply with the requirements for a conditioned space. Either the interior or exterior envelope of unconditioned vestibules shall comply with the requirements for a conditioned space

Options

EN19 Alternative Constructions (Climate Zones: all)

The climate zone recommendations provide only one solution for upgrading the thermal performance of the envelope. Other constructions can be equally effective, but they are not shown in this document. Any alternative construction that is less than or equal to the U-factor, C-factor, or F-factor for the appropriate climate zone construction is equally acceptable. A table of U-factors, C-factors, and F-factors that correspond to all the recommendations is presented in Appendix A.

Procedures to calculate U-factors and C-factors are presented in *ASHRAE Handbook—Fundamentals* (ASHRAE 2009), and expanded U-factor, C-factor, and F-factor tables are presented in Appendix A of ASHRAE/IES Standard 90.1 (ASHRAE 2010a).

Cautions

The design of building envelopes for durability, indoor environmental quality, and energy conservation should not create conditions of accelerated deterioration or reduced thermal performance or problems associated with moisture, air infiltration, or termites.

The following cautions should be incorporated into the design and construction of the building.

EN20 Truss Heel Heights (Climate Zones: all)

When insulation levels are increased in attic spaces, the truss heel height should be raised to avoid or at least minimize the eave compression. Roof insulation should extend to the exterior of the walls to minimize edge effects.

EN21 Moisture Control (Climate Zones: all)

Building envelope assemblies (see Figure 5-13) should be designed to prevent wetting, high moisture content, liquid water intrusion, and condensation caused by diffusion of water vapor. See Chapter 24 of *ASHRAE Handbook—Fundamentals* (ASHRAE 2009) for additional information.

EN22 Thermal Bridging—Opaque Components (Climate Zones: all)

Thermal bridging in opaque components occurs when continuous conductive elements connect internal and external surfaces. The adverse effects of thermal bridging are most notable in cold climates where frost can develop on internal surfaces and lead to water droplets when

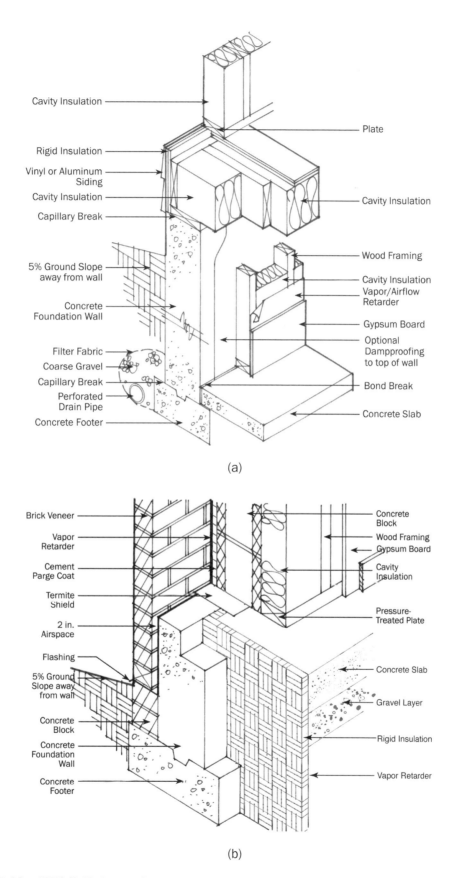

Figure 5-13 (EN21) Moisture Control for (a) Mixed Climates and (b) Warm, Humid Climates

the indoor temperature increases. The solution to thermal bridging is to provide thermal breaks or continuous insulation. Common problem areas are parapets, foundations, and penetrations of insulation.

The thermal bridge at parapets is shown in Figure 5-14a. The problem is that a portion of the wall construction is extended to create a parapet that extends above the roof to ensure worker safety per local code requirements. Since the wall insulation is on the outer face of the structure, it does not naturally connect to the insulation at the roof structure. One solution is to wrap the parapet with c.i. in the appropriate locations as shown in Figure 5-14b; a structural solution is to have an independent parapet structure that periodically penetrates the roof insulation line to limit the thermal bridging effects.

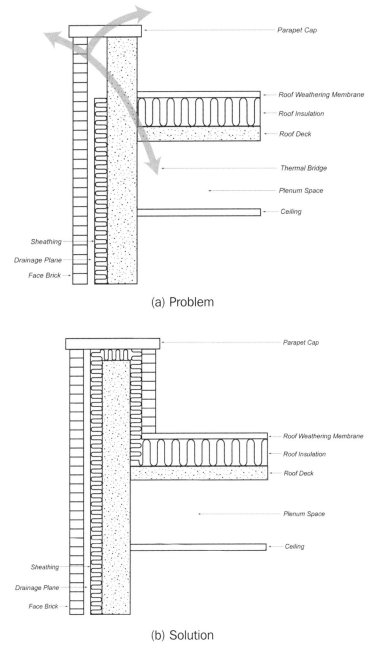

(a) Problem

(b) Solution

Figure 5-14   (EN22) Thermal Bridges at Parapets

The thermal bridge in foundations is shown in Figure 5-15a. This detail usually occurs because of construction sequences for the installation of below-grade works early in the design process. It is often an oversight to complete the connection between the below-grade and above-grade thermal protections because the installations of these elements are separate both in discipline and in time period on site. Design and construction teams must make it clear that action to establish thermal continuity of the insulation line is a performance requirement of both parties in order to achieve a typical solution as shown in Figure 5-15b. The insulation above grade needs to be protected with a surface or coating that is weather resistant and abuse tolerant.

Penetrations of insulation in which metal structural members must protrude from the building in order to support an external shade or construction (balcony, signage, etc.) need to

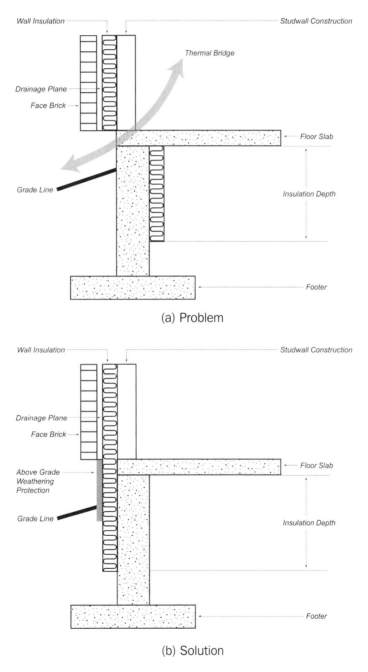

(a) Problem

(b) Solution

Figure 5-15   (EN22) Thermal Bridges at Foundations

be insulated. In these cases, the insulation should wrap the protruding metal piece when it is within the indoor cavity, and an additional length of insulation should be provided on its connection in each direction in order to prevent excessive heat transfer from the metal into the internal wall cavity. It should be noted that a façade consultant can model these types of situations to advise about the various lengths and thicknesses of insulation that would be needed to limit adverse impacts from condensation within the wall cavity.

EN23   *Thermal Bridging—Fenestration* (**Climate Zones: all**)

In colder climates, it is essential to select a glazing unit to avoid large amounts of condensation. This requires an analysis to determine internal surface temperatures, since glass is a higher thermal conductor as compared to the adjacent wall in which it is mounted. There is a risk of condensation occurring on the inner face of the glass whenever the inner surface temperature approaches the room dew-point temperature.

Careful specification is also necessary to ensure that the framing of the glazed units also incorporates a thermal break.

A typical fenestration situation where thermal bridging arises is at the detailing of how a piece of well-insulated glazing abuts the opaque façade, whether it be through a metal mullion system or whether it just frames into the wall. Windows that are installed out of the plane of the wall insulation are an example of this construction (shown in Figure 5-16a). Installing the fenestration outside of the plane of the wall insulation defeats the thermal break in the window frame. In cold climates this causes condensation and frosting. The normal solution is not to rebuild the wall but to blow hot air against the window to increase the interior surface temperature of the frame and glazing, which increases the temperature difference across the glazing and reduces the interior film coefficient thermal resistance from 0.68 to 0.25 h·ft^2·°F/Btu.

Fenestration should be installed to align the frame thermal break with the wall thermal barrier (see Figure 5-16b). This minimizes the thermal bridging of the frame due to fenestration projecting beyond the insulating layers in the wall.

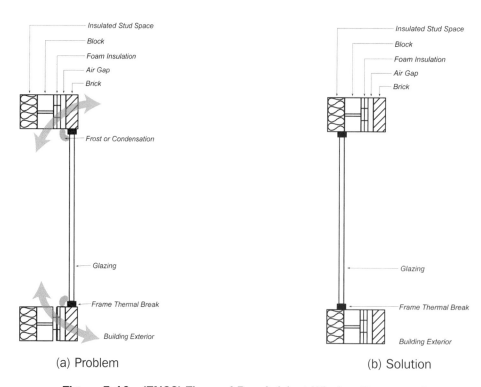

(a) Problem   (b) Solution

**Figure 5-16   (EN23) Thermal Break (a) at Window Frame and
(b) in Window Frame Aligned with Wall Insulation**

VERTICAL FENESTRATION

Good Design Practice

EN24   *Vertical Fenestration Descriptions* (Climate Zones: all)

Fenestration refers to the light-transmitting areas of a wall or roof, mainly windows and skylights but also including glass doors, glass block walls, and translucent plastic panels. Vertical fenestration includes sloped glazing if it has a slope equal to or more than 60° from the horizontal. If it slopes less than 60° from the horizontal, the fenestration falls in the skylight category. This means clerestories, roof monitors, and other such fenestration fall in the vertical category.

The recommendations for vertical fenestration are listed in Chapter 4 by climate zone. To be useful and consistent, the U-factors for windows should be measured over the entire window assembly, not just the center of glass. Look for a label that denotes the window rating is certified by the National Fenestration Rating Council (NFRC). The selection of high-performance window products should be considered separately for each orientation of the building and for daylighting and viewing functions.

The vertical fenestration descriptions listed in Table 5-2 are representative of types of windows that could be used to meet the recommended U-factors and solar heat gain coefficient (SHGC) values in the recommendation tables in Chapter 4.

To meet the SHGC recommendations for vertical fenestration in Chapter 4, use the SHGC multipliers for permanent projections as provided in Table 5.5.4.4.1 of ASHRAE/IES Standard 90.1-2010 (ASHRAE 2010a). These multipliers allow for a higher SHGC for vertical fenestration with overhangs.

EN25   *Window-to-Wall Ratio (WWR)* (Climate Zones: all)

The window-to-wall ratio (WWR) is the percentage resulting from dividing the total glazed area of the building by the total exterior wall area. For any given WWR selected

Table 5-2   Vertical Fenestration Descriptions

CZ	U-Factor	SHGC	VT	Glass and Coating	Gas	Spacer	Frame
1-3	0.46	0.23	0.51	Double clear, highly selective low-e coating		Standard	Broken aluminum
1-3	0.47	0.24	0.32	Double clear, low-e reflective coating		Standard	Broken aluminum
1-3	0.32	0.20	0.29	Double clear, low-e reflective coating		Standard	Foam-filled vinyl or pultruded fiberglass
4-5	0.34	0.25	0.51	Double clear, highly selective low-e coating	Argon	Insulated	Broken aluminum
4-5	0.35	0.22	0.32	Double clear, low-e reflective coating	Argon	Insulated	Broken aluminum
4-5	0.32	0.20	0.29	Double clear, low-e reflective coating		Standard	Foam-filled vinyl or pultruded fiberglass
6-7	0.31	0.39	0.50	Triple clear, low-e coating for outer light only	Argon	Insulated	Broken aluminum
6-7	0.26	0.31	0.54	Double clear, low-e selective coating	Argon	Insulated	Foam-filled vinyl or pultruded fiberglass
8	0.25	0.39	0.53	Triple clear, low-e coating for outer and second lights	Argon	Insulated	Aluminum thermally isolated frame
8	0.22	0.36	0.53	Triple clear, low-e coating for outer light only	Argon fill both spaces	Insulated	Foam-filled vinyl or pultruded fiberglass

CZ = climate zone

between 20% and 40%, the recommended values for U-factor and SHGC contribute toward the 50% savings target of the entire building. A reduction in the overall WWR will also save energy, especially if glazing is significantly reduced on the east and west façades. Reducing glazing on east and west façades for energy reduction should be done while maintaining consistency with regard to needs for view, daylighting, and passive solar strategies.

WINDOW DESIGN GUIDELINES FOR THERMAL CONDITIONS

Uncontrolled solar heat gain is a major cause of energy use for cooling in warmer climates and thermal discomfort for occupants. Appropriate configuration of windows according to the orientation of the wall on which they are placed can significantly reduce these problems.

EN26   *Unwanted Solar Heat Gain is Most Effectively Controlled on the Outside of the Building* (Climate Zones: all)

Significantly greater energy savings are realized when sun penetration is blocked before it enters the windows. Horizontal overhangs at the top of the windows are most effective for south-facing façades and must continue beyond the width of the windows to adequately shade them (see Figure 5-17). Vertical fins oriented slightly north are most effective for east- and west-facing facades. Consider louvered or perforated sun control devices, especially in primarily overcast and colder climates, to prevent a totally dark appearance in those environments. See DL12 for more information on shading strategies.

EN27   *Operable versus Fixed Windows* (Climate Zones: ❷B ❸ ④ ⑤ ⑥ ❼ ❽)

Operable windows play a significant role in embracing the core idea of indoor to outdoor connection and reaching out to the outdoor environment and nature. Compared to buildings with fixed-position windows, buildings with well-designed operable window systems can provide energy conservation advantages in office buildings if occupants understand their appropriate use.

However, although operable windows offer the advantage of personal comfort control and beneficial connections to the environment, individual operation of the windows not in coordination with the heating, ventilating, and air-conditioning (HVAC) system settings and require-

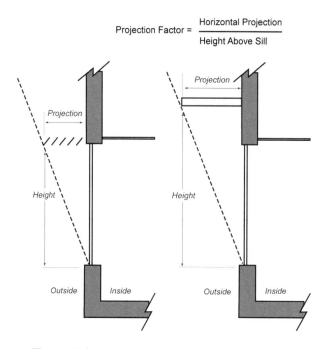

Figure 5-17 (EN26) Windows with Overhang

ments can have extreme impacts on the energy use of a building's system. Advanced-energy buildings with operable windows should strive for a high level of integration between envelope and HVAC system design. Mechanical systems should be shut off when windows are opened.

First, the envelope should be designed to take advantage of natural ventilation with well-placed operable openings. A bottom window and a top window should be opened at the same time. This allows the stack effect to set up a convection current of airflow when the difference between the indoor and outdoor temperatures is 10°F or more. Operable window systems can be controlled manually or by button-based interlock through a centralized building management system. It should be noted that ASHRAE Standard 62.1 (ASHRAE 2010b) requires that occupants have control over openings if they are used for natural ventilation (i.e., OA intake), and the adaptive comfort model in ASHRAE Standard 55 (ASHRAE 2010c) requires that occupants have continuous control over the openings if they are used as part of a natural conditioning system.

Second, the mechanical system should use interlocks and window switches on operable windows to ensure that the HVAC system responds by shutting down in the affected zone if the window is opened. The window interlock zones need to be designed to correspond as closely as possible to the HVAC zones affected by the open windows.

Third, in some cases open windows at night may be used to remove thermal loads that have accumulated over the course of the day. Occupancy types best suited to this night flush or night purge include areas with no occupancy during the cool evening and early morning hours. To allow for this to happen, the following conditions are required:

- Footprint: narrow floor plate and open-plan layout
- Operable windows controlled by a building management system that tracks the temperature of the slab, the temperature of the outdoors, and the external and internal relative humidities
- Solid slabs/exposed ceiling slabs in concrete structures
- Minimal concerns over building security risks due to open windows

Warm Climates

EN28 *Building Form and Window Orientation* (Climate Zones: ❶ ❷ ❸)

In warm climates, south-facing glass can be more easily shielded and can result in less solar heat gain and glare than can east- and west-facing glass. During early building configuration studies and predesign, preference should be given to site layouts that permit elongating the building in the east-west direction and that permit orienting more windows to the north and south. A good design strategy avoids areas of glass that do not contribute to the view from the building or to the daylighting of the space. If possible, configure the building to maximize north- and south-facing walls and glass by elongating the floor plan. Since sun control devices are less effective on the east and west façades, the solar penetration through the east- and west-facing glazing should be minimized. This can be done by reducing the area of glazing or, if the glass is needed for view or egress, by reducing the SHGC, or by utilizing automated operable shading systems. For buildings where a predominantly east-west exposure is unavoidable, more aggressive energy conservation measures will be required in other building components to achieve an overall 50% energy savings. See DL5 and DL6 for more information on building orientation and shape as they relate to daylighting strategies.

EN29 *Glazing* (Climate Zones: ❶ ❷ ❸)

For north- and south-facing windows, select windows with low SHGCs and an appropriate visible transmittance (VT); see EN33. Certain window coatings, called *selective low-e*, transmit the visible portions of the solar spectrum selectively, rejecting the nonvisible infrared sections. These glass and coating selections can provide a balance between VT and solar heat gain. Window manufacturers market special "solar low-e" windows for warm climates. All values are for the entire fenestration assembly, in compliance with NFRC procedures, and are not simply center-of-glass values. For warm climates, a low SHGC is much more important for low energy use than the window assembly U-factor. Windows with low SHGC values will tend to have a low center-of-glass U-factor because they are designed to reduce the conduction of the solar heat gain absorbed on the outer layer of glass through to the inside of the window.

EN30   *Obstructions and Plantings* (Climate Zones: all)

Adjacent taller buildings and trees, shrubs, or other plantings effectively shade glass on south, east, and west facades. For south-facing windows, remember that the sun is higher in the sky during the summer, so shading plants should be located high above the windows to effectively shade the glass. Also, be careful to not block south light that is being counted on for daylighting. While the shading effect of plants can reduce energy consumption, it doesn't impact equipment size. The sizing of HVAC equipment relies on the SHGC of the glass and shading system only. The glazing of fully shaded windows can be selected with higher SHGC ratings without increasing energy use.

The solar reflections from adjacent buildings with reflective surfaces (metal, windows, or especially reflective curtain walls) should be considered in the design. Such reflections may modify shading strategies, especially on the north facade.

Cold Climates

EN31   *Window Orientation* (Climate Zones: ④ ⑤ ⑥ ❼ ❽)

Only the south glass receives much sunlight during the cold winter months. If possible, maximize south-facing windows by elongating the floor plan in the east-west direction and relocate windows to the south face. Careful configuration of overhangs or other simple solar control devices will allow for passive heating when desired but prevent unwanted glare and solar overheating in the warmer months. To improve performance, operable shading systems should be employed that achieve superior daylight harvesting and passive solar gains and also operate more effectively when facing east and west directions. Unless such operable shading systems are used, glass facing east and west should be significantly limited. Areas of glazing facing north should be optimized for daylighting and view and focus on low U-factors to minimize heat loss and maintain thermal comfort by considering triple glazing to eliminate drafts and discomfort. During early building configuration studies and predesign, preference should be given to sites that permit elongating the building in the east-west direction and that permit orienting more windows to the south. See DL5 and DL6 for more information on building orientation and shape as they relate to daylighting strategies.

EN32   *Passive Solar* (Climate Zones: ④ ⑤ ⑥ ❼ ❽)

Passive solar energy-saving strategies should be limited to nonpermanently occupied spaces such as lobbies and circulation areas, unless those strategies are designed so that the occupants are not affected by direct beam radiation. Consider light-colored blinds, blinds within the fenestration, light shelves, or silk screen ceramic coating (frit) to control solar heat gain. In spaces where glare is not an issue, the usefulness of the solar heat gain collected by these windows can be increased by using hard massive and darker-colored floor surfaces such as tile or concrete in the locations where the transmitted sunlight will fall. These floor surfaces absorb the transmitted solar heat gain and release it slowly over time, providing a more gradual heating of the structure. Consider higher SHGC and low-e glazing with optimally designed exterior overhangs.

EN33   *Glazing* (Climate Zones: ④ ⑤ ⑥ ❼ ❽)

Higher SHGCs are allowed in colder regions, but continuous horizontal overhangs are still necessary to block the high summer sun angles.

WINDOW DESIGN GUIDELINES FOR DAYLIGHTING

Good Design Practice

EN34 *Visible Transmittance (VT)* (Climate Zones: all)

Using daylight in place of electrical lighting significantly reduces the internal loads and saves cost on lighting and cooling power. In the U.S., it is estimated that 10% of the total energy generated in 24 hours is consumed by electrical lighting during daytime. The higher the VT, the more energy that can be saved.

The amount of light transmitted in the visible range affects the view through the window, glare, and daylight harvesting. For the effective use of daylight, high-VT glazing types (0.60 to 0.70) should be used in all occupied spaces.

High VT values are preferred in predominantly overcast climates. VT values below 0.50 appear noticeably tinted and dim to occupants and may degrade luminous quality. However, lower VT values may be required to prevent glare, especially on the east and west façades or for higher WWRs. Lower VT values may also be appropriate for other conditions of low sun angles or light-colored ground cover (such as snow or sand), but adjustable blinds should be used to handle intermittent glare conditions that are variable.

High continuous windows are more effective than individual ("punched") or vertical slot windows for distributing light deeper into the space and provide greater visual comfort for the occupants. Try to expand the tops of windows to the ceiling line for daylighting, but locate the bottoms of windows no higher than 30 in. above the floor (for view). Daylighting can be achieved with higher WWRs, which can lead to higher heating and cooling loads.

EN35 *Separating Views and Daylight* (Climate Zones: all)

In some cases, daylight harvesting and glare control are not always best served by the same glazing product. Perimeter zones in particular require better control of visual comfort levels, which can make it necessary to separate daylight glazing from view glazing.

The most common strategy is to separate (split) the window horizontally to maximize daylight penetration. For daylight glazing, which is located above the view window, between 6 ft above the floor and the ceiling, high-VT glazing should be used. The view windows located below 6 ft do not require such high VT values, so values between 0.50 and 0.60 are acceptable to achieve recommended SHGC values. See DL7–DL12 for more information on vertical glazing strategies.

Windows both for view and for daylighting should primarily be located on the north and south façades. Windows on the east and west should be minimized, as they are difficult to protect from overheating and from glare. See DL4–DL6 for more information on building orientation and layout in regards to daylighting.

EN36 *Color-Neutral Glazing* (Climate Zones: all)

The desirable color qualities of daylighting are best transmitted by spectrally neutral glass types that alter the color spectrum the smallest possible extent. Avoid tinted glass, in particular bronze- and green-tinted glazing.

EN37 *Reflectivity of Glass* (Climate Zones: all)

To the greatest extent possible, avoid the use of reflective glass or low-e coatings with a highly reflective component. These reduce transparency significantly, especially at acute viewing angles, where they impact the quality of the view.

EN38 *Light-to-Solar-Gain Ratio* (Climate Zones: all)

High-performance and selective low-e glazing permit significantly higher VT than reflective coatings or tints. The light-to-solar-gain ratio is the criterion for stating the efficacy of the glass, indicating the ability to maximize daylight and views while minimizing solar heat gain. In today's markets, a variety of cost-effective glass types are available with high light-to-solar-gain ratios. Ratios over 1.6 are considered good. Any ratio greater than 2.0 is very effective and will contribute to achieving the goal of 50% energy savings.

EN39   ***High Ceilings* (Climate Zones: all)**

More daylight savings will be realized if ceiling heights are raised along the building perimeter. Greater daylight savings can be achieved by increasing ceiling heights to 11 ft or higher and by specifying higher VT values (0.60 to 0.70) for the daylight windows than for the view windows. North-facing clerestories are more effective than skylights to bring daylight into the building interior.

EN40   ***Light Shelves* (Climate Zones: all)**

Consider using interior or exterior light shelves between the daylight windows and the view windows. These are effective for achieving greater uniformity of daylighting and for extending ambient levels of light onto the ceiling and deeper into the space. Some expertise and analysis will be required to design an effective light shelf.

REFERENCES

ASHRAE. 2009. *ASHRAE Handbook—Fundamentals*. Atlanta: American Society of Heating, Refrigerating and Air-Conditioning Engineers.

ASHRAE. 2010a. ANSI/ASHRAE/IES Standard 90.1-2010, *Energy Standard for Buildings Except Low-Rise Residential Buildings*. Atlanta: American Society of Heating, Refrigerating and Air-Conditioning Engineers.

ASHRAE. 2010b. ANSI/ASHRAE Standard 62.1-2010, *Ventilation for Acceptable Indoor Air Quality*. Atlanta: American Society of Heating, Refrigerating and Air-Conditioning Engineers.

ASHRAE. 2010c. ANSI/ASHRAE Standard 55-2010, *Thermal Environmental Conditions for Human Occupancy*. Atlanta: American Society of Heating, Refrigerating and Air-Conditioning Engineers.

ASTM. 2011. ASTM E1980-11, *Standard Practice for Calculating Solar Reflectance Index of Horizontal and Low-Sloped Opaque Surfaces*. West Conshohocken, PA: ASTM International.

ASTM. 2003. ASTM E2178-03, *Standard Test Method for Air Permeance of Building Materials*. West Conshohocken, PA: ASTM International.

DAYLIGHTING

Daylighting is based on an integrated approach to design that influences the building at every scale and level of design and during each phase of the design process. Multiple design approaches and technologies can be applied to a building project when incorporating daylighting.

GENERAL RECOMMENDATIONS

When considering daylighting in a building design (see Figure 5-18), consider the following actions.

- Use a shallow floor plate so that all employees are within 30 ft of perimeter windows for daylight and views.
- Locate open-plan workstations next to windows and use low partitions with translucent materials so daylight penetrates deep into the building.
- Place corridors between open-plan workstations and private offices to use spill light to supplement electric lights in corridors.
- Locate private offices on east and west perimeters and interiors at the boundaries of daylight zones with glazing parallel to perimeter walls.
- When not located along window walls, build conference rooms with glass walls parallel to perimeter walls.
- Use light-colored matte finishes to promote interreflections and better utilization of electric light and daylight.
- Use local articulated task lights to supplement daylight and electric light.
- For controls, include daylight dimming in open offices, time switches (time clocks) in corridors, and vacancy sensing in private offices and conference rooms.

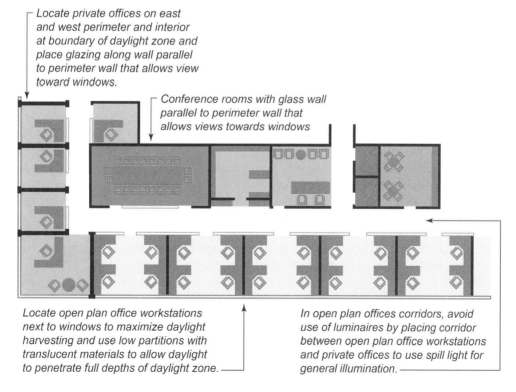

Locate private offices on east and west perimeter and interior at boundary of daylight zone and place glazing along wall parallel to perimeter wall that allows view toward windows.

Conference rooms with glass wall parallel to perimeter wall that allows views towards windows

Locate open plan office workstations next to windows to maximize daylight harvesting and use low partitions with translucent materials to allow daylight to penetrate full depths of daylight zone.

In open plan offices corridors, avoid use of luminaires by placing corridor between open plan office workstations and private offices to use spill light for general illumination.

Figure 5-18   Daylighting Design Concepts

DL1 *Daylighting Early in the Design Process* (Climate Zones: all)

In office buildings, the program and site plan are the main drivers that establish the shape and the footprint of the building. Planning criteria often result in creating compact, deep floor plates, while daylighting strategies attempt the opposite by articulating and narrowing the floor plate.

The configuration of the building footprint is established early in the design process, freezing the building depth early and locking in all future potential for daylighting. The building footprint is the key factor for anticipating future design upgrades and improvements. A frequent issue with existing buildings is their depth of floor plate, which prevents easy upgrades with daylighting and natural ventilation.

This demonstrates two important aspects. One is the importance of integrating daylight design criteria before the footprint is locked in so that the building can unfold its full energy-saving potential. Another is that space planning and energy-efficient design are inseparable design criteria, as they both impact the shape and footprint and are integral drivers of the shape of the building.

Daylight strategies impact the design at different levels of scale in each phase of design and can be characterized in four categories.

Predesign. During predesign, the daylight strategies' focus is on building configuration studies and the shaping of the floor plate. The goal is to minimize depth and maximize access to windows and daylight by strategically placing light wells, shafts, and atriums and orienting fenestration in a predominantly north- and south-facing direction. The emphasis is on maximizing the amount of occupied space that has access to windows and on minimizing the distance from the building core to the perimeter.

Schematic Design. During the schematic design phase, daylight strategies are about interiors, focusing on spatial considerations to optimize daylight penetration and defining ceiling height, layout, and partition wall transparency with clerestory windows for borrowed light. The planning focus is directed toward coordinating space types that require daylight and views and placing them along the perimeter.

Design Development. During the design development phase, the daylighting strategies' focus is on envelope design to optimize quantity and quality of daylight while minimizing solar gains. The interior design focus is on surface reflectivity and optimizing furniture and partition layout to align with visual and thermal comfort requirements.

Construction Documents (CDs). Coordination of electrical lighting includes the placement of photosensors and occupancy sensors for controlling automated daylight switching and dimmable ballasts.

DL2 *Daylighting Analysis Tools to Optimize Design* (Climate Zones: all)

This Guide is designed to help achieve energy savings of 50% without energy modeling, but energy and daylighting modeling programs make evaluating energy-saving trade-offs faster and daylighting designs far more precise.

Annual savings will have to be calculated with an annual whole-building energy simulation tool after the daylighting design tools have been used to determine the footcandles (fc) in the spaces and after the windows have been appropriately sized. Current daylighting analysis tools do not help with heating and cooling loads or other energy uses; they predict only illumination levels and electric lighting use.

DL3 *Space Types, Layout, and Daylight* (Climate Zones: all)

In office buildings, daylight is a key requirement for all regularly occupied spaces. Daylighting is essential for worker performance. It is a design strategy that dovetails with the effort to save energy through reduction of electric light and cooling loads.

The goal is to identify the spaces that best lend themselves to daylight harvesting and saving energy and to recommend layout strategies that allow locating spaces on the perimeter of the building. The potential of energy saving through daylighting varies and depends on program and space types, which can be broadly characterized by the following four categories of occupied spaces.

Open-Plan Offices. These spaces are high-density populated spaces. From an energy performance standpoint, the first priority is to locate open office areas, which are the most beneficial spaces for harvesting daylight, on the perimeter, preferably in a north- and south-facing configuration. The following tips apply to open office spaces.

- Locate workstations next to windows within the primary and secondary daylight zones to maximize daylight harvesting.
- Use low partitions with translucent materials to allow daylight to penetrate the full depth of the primary and secondary daylight zones.
- Use local articulated task lights to supplement daylight and electric light.
- Use appropriate combinations of controls, including daylight sensing, time clocks, and vacancy sensing.

Open-Plan Office Corridors. Avoid the use of luminaires in open-plan office corridors by utilizing spill light from electrical lighting and daylighting. Locate corridors along the boundary of the perimeter daylight zone between the open-plan office workstations and inbound private offices or conference rooms.

Private Offices. These spaces are low-density populated spaces. Locate private office spaces on the east- and west-facing perimeters or on the interior at the boundary of the secondary daylight zone. For private offices not on the window wall, place glazing along the private office wall that is parallel to perimeter wall, which allows views toward windows. Control by manual ON single-zone occupancy sensors or automatic ON to 50% two-zone occupancy sensors.

Conference Rooms. Conference rooms are low-density populated spaces that build up high interior heat loads for a limited period of time. Locate conference rooms on the interior at the boundary of the secondary daylight zone. Use translucent or transparent glass walls that exchange light with adjacent corridors and offices. Place glazing along the wall that is parallel to the perimeter wall, which allows views toward windows. When located on the perimeter, the interior loads and solar radiation penetrating the perimeter wall accumulate, leading to escalation of peak loads and oversizing of HVAC systems. As a strategy to minimize peak loads, conference rooms should be located on north façade perimeters only or inboard, avoiding west-, south-, and east-facing perimeter walls. This approach is supported by prioritizing perimeter space for permanently occupied spaces, which make better use of daylight and views than conference rooms, which remain unoccupied in many cases.

Public Spaces (Lobbies, Reception Areas, Waiting Areas, and Transitional Spaces). These spaces provide the best opportunity for high ceilings with high, large-scale fenestration and offer large potential for daylight harvesting and energy savings due to their depth and potentially high ceilings.

The following recommendations apply to spaces that are not located on the building perimeter but will allow for additional energy savings if they are designed to follow specific rules.

Internal Corridors. In single-story buildings or on top-level floors, where sidelighting is not available, toplighting should be used to provide daylight for corridors and contiguous spaces. Make sure that reception areas, which are frequently placed in niches of circulation areas, have access to daylight and views.

Daylight Zone Definitions

The *primary daylight zone* depth extends one window head height into the space (head height is the distance from the floor to the top of the glazing), and the width of the daylight zone is the width of the window plus two feet on each side. The *secondary daylight zone* extends from the end of the primary daylight zone an additional head height into the space. Either daylight zone ends at a 5 ft or higher vertical partition.

Reception and Pantry Areas. These spaces are low-density populated spaces. Use automatic time controls that switch lights on during normal hours and off after hours but allow manual override.

DL4   *Building Orientation and Daylight* (Climate Zones: all)

Effective daylighting begins with selecting the correct solar orientation of the building and the building's exterior spaces. For most spaces, the vertical façades that provide daylighting should be oriented within 15° of north and south directions. Sidelighted daylighting solutions can also work successfully for other orientations, but they will require a more sophisticated approach to shading solutions, and they would reach beyond the recommendations proposed for accomplishing the goals stipulated in this Guide.

Context and Site. Ensure that apertures are not shaded by adjacent buildings, trees, or components of the office building itself.

DL5   *Building Shape and Daylight* (Climate Zones: all)

The best daylighting results are achieved by limiting the depth of the floor plate and minimizing the distance between the exterior wall and any interior space. Narrowing the floor plate will in most cases result in introducing courtyards and articulating the footprint for better daylight penetration.

Building Shape and Self-Shading. Optimizing the building shape for daylight translates to balancing the exterior surface exposed to daylight and self-shading the building mass to avoid direct-beam radiation. Items to consider include the following.

• Locate the maximum amount of occupied within minimum distance to the building perimeter for effective daylighting.

• Shape the building footprint and plan fenestration so that all occupants are within 30 ft of perimeter fenestration.

• Shape the building footprint to allow for all regularly occupied spaces within 15 ft of the perimeter to be equal to or exceed 40% of the total floor plate area.

• Ensure that 75% of the occupied space is located within 20 ft of the perimeter wall.

• Target the floor plate to achieve a depth of 60 ft where toplighting is not an option.

For sunny climates, designs can be evaluated on a sunny day at the summer solar peak. For overcast climates, a typical overcast day should be used to evaluate the system. Typically, the glazing-to-floor ratio percentage will increase for overcast climates. Daylighting can still work for an office building in a overcast climate; however, overcast climates often produce diffuse skies, which create good daylighting conditions and minimize glare and heat gain.

Daylighting systems need to provide the correct lighting levels. To meet the criteria, daylight modeling and simulation may be required. Daylighting systems should be designed to meet the following criteria.

• In a clear sky condition, to provide sufficient daylight, illuminance levels should achieve a minimum of 25 fc but no more than 250 fc.

• In overcast conditions, daylighted spaces should achieve a daylight factor of 2% but not exceed a daylight factor of 20%.

The same criteria for lighting quality and quantity apply to electric lighting and daylighting. When the criteria cannot be met with daylighting, electric lighting will meet the illuminance design criteria. The objectives are to maximize the daylighting and to minimize the electric lighting. To maximize the daylighting without oversizing the fenestration, in-depth analysis may be required.

DL6 *Window-to-Wall Ratio (WWR)* **(Climate Zones: all)**

There are two steps to approaching window configuration and sizing. The first is that the fenestration design should follow interior-driven design criteria such as occupancy type and requirements for view, daylight, and outdoor connectivity. The second step targets peak load and energy use, which limit window size to comply with the mechanical systems target. For office buildings to achieve 50% savings, the overall WWR should not exceed 40%.

DL7 *Sidelighting—Ceiling and Window Height* **(Climate Zones: all)**

For good daylighting in office-type spaces, a minimum ceiling height of 9 ft is recommended. In public spaces and lobbies that extend to greater depth, ceiling height, at least partially, should be 10 to 12 ft. When daylighting is provided exclusively through sidelighting, it is important to elevate the ceiling on the perimeter and extend glazing to the ceiling. Additional reflectance to increase lighting levels can be achieved by sloping the ceiling up toward the outside wall. (See Figure 5-19.)

The effective aperture (EA) for sidelighting is the area of glazing in an unobstructed wall multiplied by the VT of vertical glazing, divided by the floor area in the primary daylight zones. The EAs in the recommendation tables in Chapter 4 of this Guide were derived from energy analysis for the Pacific Northwest National Laboratory (PNNL) *Technical Support Document: 50% Energy Savings Design Technology Packages for Medium Office Buildings* (Thornton, et al. 2009).

DL8 *Sidelighting—Clerestory Windows* **(Climate Zones: all)**

In cases where it is not possible to place windows in exterior walls for programmatic or functional reasons, clerestory windows or window bands should be considered for daylighting. Daylight delivered above 7 ft, at clerestory level, delivers the highest illuminance level available through sidelighting. (See Figure 5-20.)

DL9 *Sidelighting—Borrowed Light* **(Climate Zones: all)**

Borrowed light is an effective strategy for delivering daylight to corridors that are located behind spaces on the building perimeter. The corridor wall frequently blocks and prevents daylight from entering deeper into the building. Partitioning the corridor wall provides significant opportunities to daylight the corridor through borrowed light. Corridor partitions should be designed with clerestory windows or window bands for perimeter spaces with a depth-to-height ratio no larger than 2.5:1. (See Figure 5-21.)

DL10 *Sidelighting—Wall-to-Wall Windows* **(Climate Zones: all)**

Raising window levels to ceiling level is the first priority for deepening daylight penetration. However, to balance light levels in the room and to mitigate contrast, it is equally important

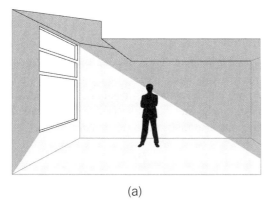

(a)

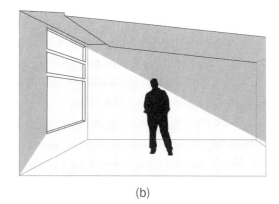

(b)

Figure 5-19 (DL7) (a) Raised Ceiling at Façade and (b) Sloped Ceiling at Façade

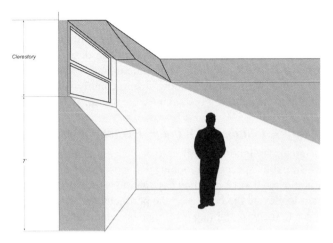

Figure 5-20   (DL8) Clerestory

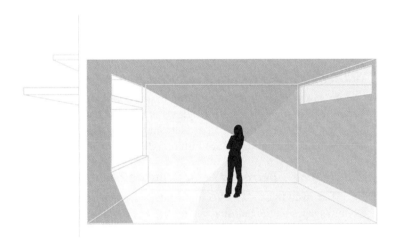

Figure 5-21   (DL9) Borrowed Light in Corridor

to maximize window width. By extending the window width from wall to wall, the adjacent partitioning walls receive greater exposure and act as indirect sources of daylight while also achieving greater depth of daylight penetration. (See Figure 5-22.)

Even more daylight and a wider range of view can be gained by making the first 2 to 3 ft of cellular partitioning walls, where they meet the perimeter wall, transparent. This enlarges the daylighted portion of the room enclosure by 50% to 60% per space. (See Figure 5-23.)

DL11   *Sidelighting—Punched Windows* (Climate Zones: all)

In cases where window size is limited and "punched" windows can't be avoided, special care should be taken in placing the apertures to avoid high contrasts and low visual comfort.

To ensure that daylight is maximized and light levels are distributed evenly, the window aperture should align with either of the partitioning walls. This will mitigate contrast differences, maximize the depth of daylight reach, and also make the space appear larger. (See Figure 5-24.)

DL12   *Shading Systems to Eliminate Direct-Beam Radiation* (Climate Zones: all)

Essential for good daylight quality in office building design is the elimination of uncontrolled direct-beam sunlight impacting workstation areas. Direct-beam radiation causes thermal

Figure 5-22   (DL10) Maximized Window Width

Figure 5-23   (DL10) Transparent Partitioning Wall

Figure 5-24   (DL11) Punched Window Placed next to Partition Wall

Protection from Direct Solar Radiation

Frosted Glass Fins
Source: Rick Lasser, Arup

Frosted glass fins at the northwest façade of the San Francisco Federal Office Building serve to protect the glass façade from direct solar radiation during the early afternoon hours, when the sun is at shallow angles with respect to the façade. Spacing and depth of the fins are tuned to cut out direct sun until after 4:00 p.m. on summer afternoons.

discomfort and glare, which are critical to avoid in all office spaces but less critical to avoid for some public spaces and corridors. Strategies should be used that bounce, redirect, and filter sunlight so that direct radiation does not enter the space.

The sun is a moving source of energy with constantly changing directions and intensities of light and heat radiation. When planning the exterior walls, designers face the task of minimizing the solar heat load but maximizing glare-free daylight under permanently changing conditions. The goal is to maximize the light-to-solar-heat-gain ratio for every minute of the day.

Shading systems are designed to reduce solar radiation. However, in most cases they also inadvertently cause loss of valuable daylight. As a result, the electric lights are switched on during the peak time of day, causing cooling load and power consumption to peak and driving HVAC sizing excessively/uncontrollably. This explains why in the process of developing a shading strategy it becomes inevitable to acknowledge and include daylighting as an integral component of the system.

The effectiveness of shading systems varies widely and depends on a system's ability to adapt to changing conditions. This explains why dynamic systems that operate on demand and track the path of the sun are significantly more successful than static/fixed systems.

Shading Type Selection. To obtain the best overall performance results, the selection of the right shading type should be based on considerations concerning both heat load and the ability to facilitate daylight and views. There are six shading types to choose from, as discussed in the following.

- *Fixed External Shading.* Solar heat gain is most effectively controlled when penetration is blocked before entering the building. One disadvantage of exterior shading systems can be

Solar Control

Double-Glazed Windows with Interstitial Blinds
Source: Gerard Healey, Arup

At the Syracuse Center of Excellence, double-glazed windows with frit and interstitial blinds provide solar control. Along the bottom of the larger, fixed-glass windows there are also smaller, operable windows to allow a mixed-mode approach.

the accessibility issues for maintaining and cleaning the façade. Fixed devices are designed to perform best at peak hours but work significantly less effectively outside the optimized time range. There are two different configurations of exterior shading:

- *Horizontal Devices.* Overhangs, soffits, awnings, and trellises respond well to steep solar angles and work best on south-facing façades. Passive solar gains are possible in winter; however, additional interior shading will be required to counter glare. A projection factor (PF) of 0.5 is typical. Overhangs are most effective and economical when they are located directly above the glass and continue beyond the width of the window. (See Figures 5-25a and 5-25b.)

- *Vertical Devices.* Vertical screens or horizontal louvers configured in vertical arrays work when oriented south, west, or east. (See Figure 5-25c.)

- *Dynamic Shading Systems.* Dynamic or operable systems are the most effective shading devices available, as they don't have to compromise on one single position for minimizing heat gain and maximizing daylight. The most common technologies used are louvered systems and fabric-based roller shades, which are able to reduce solar heat gain by as much as 80% to 90% while concurrently allowing for daylight and views. Operable systems are motorized and controlled either manually or automatically and can be driven by solar tracking technology.

- *Exterior Systems.* On the exterior, operable systems are less commonly used than fixed systems due to higher maintenance and vulnerability in windy conditions. The best applications are found in double-skin façade systems, a rapidly emerging technology, where accessibility is easier and weather protection allows for more lightweight solutions.

- *Interstitial Systems.* Operable louvers are located between the glass panes in an integrated insulated glazing unit. Louvers are rotated, raised, and lowered electrically. This clean, tidy solution with good accessibility lends itself to application in office environments. (See Figure 5-26.)

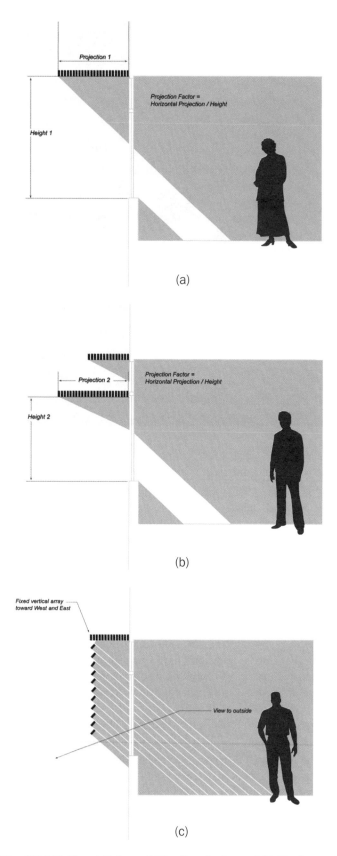

Figure 5-25   (DL12) Fixed External Shading in (a and b) Horizontal Configuration and (c) Vertical Configuration

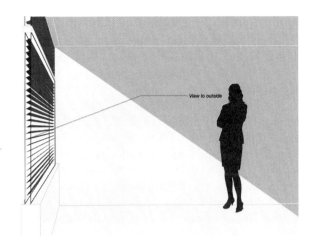

Figure 5-26   (DL12) Operable Louvers Located between Glass Panes

- *Internal Shading Systems.* Fixed interior shades or operable roller shade systems are typically used for filtering light to mitigate glare or to ensure thermal comfort against direct solar radiation that has escaped the exterior shading devices. Internal shades use fabrics with various combinations of SHGCs, openness factors, colors, and weaves to reduce heat radiation and provide thermal comfort and glare protection.

    Using internal shading alone, without automated daylight control systems, should not be a primary strategy for improving energy performance. Internal shading systems can, however, be used for improving visual comfort. Interior light shelves can also act as internal shades, but their increased requirement for cleaning and maintenance requires serious consideration.

- *Automated or Manual Operation.* The success of daylighted spaces depends on how occupants interact with the daylighting system. This is particularly true for blinds or shades that are available for adjustment by occupants. Occupants are motivated to close the blinds but not to reopen them. If blinds are left closed, the daylighting potential will not be realized. Automated systems allow user override but can be programmed to reset themselves to their system/performance-based position.

    Also, if temporary darkening of a specific space is not functionally required, do not install shades or blinds on the daylighting glass. Unnecessary blinds will result in reduced performance, increased first costs, and higher long-term maintenance expenses.

DL13   Daylighting Control for Audiovisual Activities (Climate Zones: all)

If a space requires darkening for audiovisual or other functions, consider motorized roller shades or motorized vertical blinds for apertures that are out of reach. This may seem as if it will result in higher maintenance costs, but such controls can have the opposite effect. The mechanical stress placed on manual operators by the personnel (because of uneven cranking) limits the effective life of these devices to fewer than 10 years. The inconvenience associated with the process also results in a number of these shades being left closed. Motorized shades, which cost more up front, provide operators with greater ease of operation and result in a better-performing daylighting design. Some motorized devices can also be programmed to reset in the open position at the beginning of each day.

Television monitors or liquid crystal display (LCD) projectors require that the light level at the specific location of the screen falls in the range of 5 to 7 fc for optimum contrast. Slightly higher levels (7 to 15 fc) should still provide acceptable light levels for the visual aids, but the reduced contrast will make them harder to read.

As an option to shading the daylighting apertures, consider locating the screen or monitor in a part of the room that has less daylight and does not produce glare on the screen.

DL14   *Interior Finishes for Daylighting* (Climate Zones: all)

Select light colors (white is best) for interior walls and ceilings to increase light reflectance and reduce lighting and daylighting requirements. Minimum surface reflectances are shown in Table 5-3. The colors of the ceiling, walls, floor, and furniture have major impacts on the effectiveness of the daylighting strategy.

Consider a ceiling tile or surface that has a high reflectivity. Make sure that the ceiling tile reflectance includes the fissures within the acoustical tiles, as these irregularities affect the amount of light absorbed. Do not assume that the color of a tile alone dictates its reflectance. When selecting a tile, specify a minimum reflectivity. Most manufacturers will list the reflectance as if it were the paint color reflectance. The commissioning (Cx) provider should verify the reflectance. See EL1 for additional information on interior finishes.

DL15   *Outdoor Surface Reflectance* (Climate Zones: all)

Consider the reflectances of the roofs, sidewalks, and other surfaces in front of the glazing areas of the building. The use of lighter roofing colors can increase daylighting concentration and in some cases can increase indoor illuminance levels to reduce power consumption for electrical lighting.

High-albedo roofs reflect heat instead of absorbing it; they help lower the heat load and keep the building cooler. Also, the heat-island effect is diminished, which lowers the environmental temperature, which can support natural ventilation through courtyard fenestration.

Use caution, however, when designing light-colored walkways in front of floor-to-ceiling glazing. Light-colored surfaces will improve daylighting but can also cause unwanted reflections and glare impacting interior spaces.

DL16   *Calibration and Commissioning* (Climate Zones: all)

Even a few days of occupancy with poorly calibrated controls can lead to permanent overriding of the system and loss of savings. All lighting controls must be calibrated and commissioned after the finishes are completed and the furnishings are in place. Most photosensors require daytime and nighttime calibration sessions. The photosensor manufacturer and the quality assurance (QA) provider should be involved in the calibration. Document the calibration and Cx settings and plan for future recalibration as part of the maintenance program.

DL17   *Dimming Controls* (Climate Zones: all)

In all regularly occupied daylighted spaces such as staff areas, continuously dim rather than switch electric lights in response to daylight to minimize occupant distraction. Specify dimming ballasts that dim to at least 20% of full output and that have the ability to turn off when daylighting provides sufficient illuminance. Provide a means and a convenient location to override daylighting controls in spaces that are intentionally darkened to use overhead projectors or slides. The daylighting control system and photosensor should include a 15-minute time delay or other means to avoid cycling caused by rapidly changing sky conditions and a 1-minute fade rate to change the light levels by dimming. Automatic multilevel daylight switching may be used in

Table 5-3 Minimum Reflectances

Location	Minimum Reflectance
Wall segment above 7 ft	70%
Ceiling	70% (preferably 80%–90%)
Light well	70%
Floor	20%
Furniture	50%
Walls segment below 7 ft	50%

environments that are not regularly occupied, such as hallways, storage rooms, restrooms, lounges, and lobbies.

DL18   *Photosensor Placement and Lighting Layout* (Climate Zones: all)

Correct photosensor placement is essential. Consult daylighting references or work with the photosensor manufacturer for proper locations. Mount the photosensors in locations that closely simulate the light levels (or can be set by being proportional to the light levels) at the work planes. Depending on the daylighting strategy, photosensor controls should be used to dim particular logical groupings of lights. Implement a lighting fixture layout and control wiring plan that complement the daylighting strategy. In sidelighted spaces, locate luminaires in rows parallel to the window wall, and wire each row separately. Because of the strong difference in light that will occur close to the window and away from the window, having this individual control by bank will help balance out the space. In a space that has a skylight, install one photosensor that controls all the perimeter lights and a second that controls all the lights within the skylight well.

DL19   *Photosensor Specifications* (Climate Zones: all)

Photosensors should be specified for the appropriate illuminance range (indoor or outdoor) and must achieve a slow, smooth, linear dimming response from the dimming ballasts.

In a *closed-loop* system, the interior photocell responds to the combination of daylight and electric light in the daylighted area. The best location for the photocell is above an unobstructed location such as the middle of the space. If using a lighting system that provides an indirect component, mount the photosensor at the same height as the luminaire or in a location that is not affected by uplight from the luminaire.

In an *open-loop* system, the photocell responds only to daylight levels but is still calibrated to the desired light level received on the work surface. The best location for the photosensor is inside the skylight well.

DL20   *Select Compatible Light Fixtures* (Climate Zones: all)

First consider the use of indirect lighting fixtures that more closely represent the same effect as daylighting. Indirect lighting spreads light over the ceiling surface, which then reflects the light to the task locations; with the ceiling as the light source, indirect lighting is more uniform and has less glare.

In addition, insist on compatibility between ballasts, lamps, and controls. Ensure that the lamps can be dimmed and that the dimming ballasts, sensors, and controls will operate as a system.

REFERENCES

IES. 2011. *The Lighting Handbook*, 10th ed. NY: Illuminating Engineering Society of North America.

Thornton, B.A., W. Wang, M.D. Lane, M.I. Rosenberg, and B. Liu. 2009. *Technical Support Document: 50% Energy Savings Design Technology Packages for Medium Office Buildings*, PNNL-19004. Richland, WA: Pacific Northwest National Laboratory.

ELECTRIC LIGHTING

INTERIOR LIGHTING

Goals for Office Lighting

The primary lighting goals for office lighting are to optimize the open office spaces for daylight integration and to provide appropriate lighting levels in the private and open office spaces while not producing a dull environment. (See Figure 5-27.)

Good Design Practice

EL1   *Savings and Occupant Acceptance* (Climate Zones: all)

When using automatic daylight harvesting controls and occupancy sensors to reduce the electric lighting when daylight is present and when the space is unoccupied, it is vitally important to commission the control systems. A good control system will be invisible to the occupants, but they should be educated on the energy-saving benefits of the system and to spot and report systems that appear to be malfunctioning.

EL2   *Space Planning—Open Offices* (Climate Zones: all)

To maximize the energy savings from daylight harvesting, the open office workstations should be located on the north and south sides of the building, and all workstations need to be within the primary and secondary daylight zones. This limits the workstations to only two deep from the window wall. Additionally, the partitions separating the workstations that are parallel to the window wall must be no taller than 36 in. or be at least 50% translucent above desk height to allow daylight to reach the second workstation.

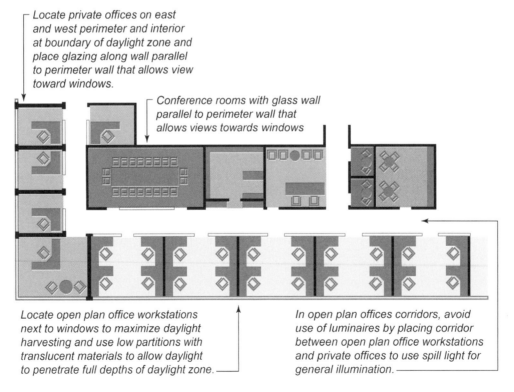

Locate private offices on east and west perimeter and interior at boundary of daylight zone and place glazing along wall parallel to perimeter wall that allows view toward windows.

Conference rooms with glass wall parallel to perimeter wall that allows views towards windows

Locate open plan office workstations next to windows to maximize daylight harvesting and use low partitions with translucent materials to allow daylight to penetrate full depths of daylight zone.

In open plan offices corridors, avoid use of luminaires by placing corridor between open plan office workstations and private offices to use spill light for general illumination.

Figure 5-27   Open Office Space Planning

EL3   Space Planning—Private Offices, Conference Rooms, and Break Rooms
(Climate Zones: all)

East- and west-facing spaces are the most difficult to daylight due to low sun angles and the tendency of tenants to close blinds. Locate private offices, conference rooms, and break rooms on the east and west sides of the building and use manual ON occupancy sensors with a daylight override to save the most energy in these spaces.

EL4   Light-Colored Interior Finishes **(Climate Zones: all)**

For electrical lighting to be used efficiently, surfaces must have light-colored finishes. Ceiling reflectance should be at least 80% (preferably 90%), which in general means using smooth white acoustical tile or ceiling paint. The average reflectance of the walls should be at least 50%, which in general means using light tints or off-white colors for the wall surfaces, as the lower reflectance of doors, tack surfaces, windows and other objects on the walls will reduce the average. Floor surfaces should be at least 20%; for this there are many suitable surfaces.

In open-plan offices, cubicle partitions should also have a reflectance of at least 50%. Partitions between cubicles that are parallel to the window wall should be at least 50% translucent or be limited to 36 in. to maximize daylight potential.

In addition, take the shape and finish of the ceiling into account. A flat painted or acoustical tile ceiling is the most efficient; sloping ceilings and exposed roof structures, even if painted white, may significantly reduce the effective ceiling reflectivity. Make sure the ceiling and all components are painted a high-reflectance white.

Reflectance values are available from paint, carpet, and ceiling tile manufacturers. Reflectance should be verified by the QA provider.

EL5   Task Lighting **(Climate Zones: all)**

If the space-planning recommendations in EL2 and EL3 are followed by locating office spaces in the daylight zones, task lighting should not be needed during the daylight hours. In daylight zones, task lights should be evaluated on a needs basis and should not be automatically installed at each workstation. If task lights are installed in the daylight zones, provide a time-clock circuit for the task lights that is set to switch the task lights off during daylight hours. Connect all task lights to plug strips that have integrated local occupancy sensors to turn the lights off when the space is unoccupied.

Periodically confirm that task lights are controlled and are turned off during daylight hours and when occupants leave the spaces during nondaylight hours.

EL6   Color Rendering Index (CRI) **(Climate Zones: all)**

The Color Rendering Index (CRI) is a scale measurement identifying a lamp's ability to adequately reveal color characteristics of objects and people. The scale maximizes at 100, with 100 indicating the best color-rendering capability. All fluorescent lamps recommended in this guide are rated at 80 CRI or greater.

EL7   Color Temperature **(Climate Zones: all)**

The color temperature is a scale identifying a lamp's relative warmth or coolness—the higher the color temperature, the bluer the source. Use either 3500K, 4100K, or 5000K fluorescent lamps.

There are preliminary studies showing that higher-color-temperature light, in the 5000K range instead of the 3500K range, may provide better visual acuity; however, 5000K lamps may produce an artificially cool-looking building at night. The higher 4100K or 5000K color temperature will also match the daylight from windows and skylights more closely than the lower 3500K color-temperature sources.

Create a purchasing plan to buy lamps in only one color temperature to maintain color consistency during spot and/or group relamping.

EL8   *Linear Fluorescent Lamps and Ballasts* **(Climate Zones: all)**

To achieve the lighting power density (LPD) recommendations in Chapter 4, high-performance T8 lamps and high-performance electronic ballasts are used for general lighting. All fluorescent lamps are temperature sensitive and produce lower light levels in cold and hot environments. What is more critical is specifying the new energy-saving T8 and T5HO lamps.

T8 High-Performance Lamps. High-performance T8 lamps are defined, for the purpose of this Guide, as having a lamp efficacy of 90+ nominal lumens per watt (LPW), based on mean lumens divided by the cataloged lamp input watts. Mean lumens are published in lamp catalogs as the reduced lumen output that occurs at 40% of the lamp's rated life. High-performance T8s also are defined as having a CRI of 81 or higher and 94% lumen maintenance. The high-performance lamp is available in 32 W rapid start and 30, 28, and 25 W instant start lamps. Table 5-4 lists the average mean LPW of the commonly manufactured 4 ft T8 lamps.

Ballasts. The ballast factor (BF) is a measure of the relative light output of the ballast. A BF of 1.0 would mean that the ballast is driving the lamp to produce 100% of the rated lamp lumens. Light output and wattage are related—the lower the BF the lower that wattage and the lower the light output. Normal BF ballasts are in the 0.85 to 1.0 range, with most at 0.87 or 0.88. Low-BF ballasts, with BFs below 0.85, can be used to reduce the light output and wattage of the system when the layout of the fixtures will overlight the space. High-BF ballasts, with BFs above 1.0, can be used to increase the light output of the lamp in areas where the fixture layout will underlight the space—wattage will go up proportionally to the BF.

Choosing Premium T8 Ballasts

NEMA Premium Ballasts
Source: Michael Lane, Lighting Design Lab

When choosing a high-performance electronic ballast, look for the National Electrical Manufacturers Association (NEMA) Premium mark on the ballast. This mark identifies ballasts that meet the Consortium for Energy Efficiency specifications for the most energy-efficient high-performance T8 ballasts available from ballast manufacturers. Generally the high-performance ballast will use 3 to 4 W less than a standard electronic ballast on a two-lamp T8 system.

Ballast efficacy factor (BEF) is a term used to compare the efficiencies of different lamp/ballast systems. BEF is [(BF · 100) / ballast input wattage]. Unfortunately, the calculated BEF changes due to the number of lamps the ballast drives, so for this Guide we modify the BEF by multiplying the calculated BEF by the number of lamps to generate a BEF-P (ballast efficacy factor—prime).

Instant Start Ballasts. High-performance electronic instant start ballasts are defined, for the purpose of this Guide, as having a BEF-P of 3.15 or greater.

For energy-saving T8 lamps, the BEF-P for 30 W systems is 3.3 or greater, for 28 W systems it is 3.6 or greater, and for 25 W systems it is 3.9 or greater.

Instant start T8 ballasts provide the greatest energy savings options and are the least costly option. Additionally, the parallel lamp operation allows one lamp to operate even if the other burns out.

Caution: Instant start ballasts may reduce lamp life when controlled by occupancy sensors or daylight switching systems. However, even if the rated lamp life is reduced by 25%, if due to the occupancy sensor the lamp is off more than 25% of the time, then the socket life (the length of time before the lamps are replaced) will be greater. If extended socket life is desired, consider program rapid start ballasts.

Program Rapid Start Ballasts. High-performance electronic program rapid start ballasts are defined, for the purpose of this Guide, as having a BEF-P of 3.00 or greater. While program rapid start ballasts are normally recommended on occupancy-sensor-controlled lamps due to increased lamp life, program rapid start ballasts use approximately 5% more power than instant start ballasts. For the recommendations in this Guide, program rapid start ballasts are not used

Table 5-4   4 ft T8 Lamp Efficacy

T8 Lamp Description	Watts	Lumens		Mean LPW	Color Temperature, K
		Initial	Mean		
F32T8/RE70	32	2800	2613	82	300, 3500, 4100
F32T8/RE70	32	2717	2535	79	5000
F32T8/RE80	32	2950	2807	88	3000, 3500, 4100
F32T8/RE80	32	2800	2660	83	5000
F32T8/RE80/HP	32	3100	2937	92	3000, 3500, 4100
F32T8/RE80/HP	32	3008	2848	89	5000
F32T8/25W/RE80	25	2458	2344	94	3000, 3500, 4100
F32T8/25W/RE80	25	2350	2241	90	5000
F32T8/28W/RE80	28	2725	2599	93	3000, 3500, 4100
F32T8/28W/RE80	28	2633	2509	90	5000
F32T8/30W/RE80	30	2850	2717	91	3000, 3500, 4100
F32T8/30W/RE80	30	2783	2653	88	5000

Note: Yellow-shaded lines indicate lamps that comply with the recommended 90+ mean lumens per watt (LPW).

BEF-P Calculation

From a lamp catalog it is known that a two-lamp ballast with 32 W lamps uses 55 W and has a BF of 0.87. With this information, calculate the BEF and BEF-P as follows.

$$BEF = (0.87 \cdot 100)/55 = 1.58$$

$$BEF\text{-}P = 1.58 \cdot 2 = 3.16$$

A BEF-P of 3.16 passes the required minimum of 3.0.

to achieve the LPDs in the recommendation tables in Chapter 4, but program rapid start ballasts may be used as long as the LPDs in Chapter 4 are not exceeded.

Caution: Using program rapid start ballasts will result in slightly higher power consumption with the same light level. The wattage and light levels will need to be reduced in other areas to meet the LPD recommendations in Chapter 4.

Dimming Ballasts. High-performance dimming ballasts are defined, for the purpose of this Guide, as having a BEF-P of 3.00 or greater. Dimming ballasts are used along with daylight controls in all open office spaces.

T5 Lamps and Ballasts. T5HO and T5 lamps have initial LPW that compare favorably to the high-performance T8s. In addition to using less energy, T5s use fewer natural resources (glass, metal, phosphors) than a comparable lumen output T8 system. However, when evaluating the lamp and ballast at the "mean lumens" of the lamps, T5HO lamps perform more poorly. On instant start ballasts, high-performance T8s are 13% more efficient than T5s. In addition, since T5s have higher surface brightness they should not be used in open-bottom fixtures. It may be difficult to achieve the LPD recommendations in Chapter 4 and maintain the desired light levels using current T5 technology as the primary light source.

EL9   *Occupancy Sensors* (Climate Zones: all)

In every application it should not be possible for the occupant to override the automatic OFF setting, even if set for manual ON. Unless otherwise recommended, factory-set occupancy sensors should be set for medium to high sensitivity with a 15-minute time delay (the optimum time to achieve energy savings without creating false OFF events). Work with the manufacturer for proper sensor placement, especially when partial-height partitions are present.

Periodically confirm that occupancy sensors are turning the lights off after the occupant leaves the space. Figure 5-28 shows a typical occupancy-sensing control setup.

Classrooms; conference, meeting, and training rooms; employee lunch and break rooms; storage and supply rooms between 50 and 1000 ft^2; rooms used for document copying and printing; office spaces up to 250 ft^2; restrooms; and dressing and locker rooms are often found in office buildings. This Guide recommends that these space types have occupancy sensors.

The greatest energy savings are achieved with manual ON/automatic OFF occupancy sensors or sensors with automatic ON to 50% light level. This avoids unnecessary operation when electric lights are not needed and greatly reduces the frequency of switching.

Caution: Confirm that the occupancy sensor is set to manual ON operation during installation. Many manufacturers ship sensors with a default setting of automatic ON.

Open Office. In open-plan offices, ceiling-mounted ultrasonic sensors should be connected to an automatic or momentary contact switch so that the operation always reverts to

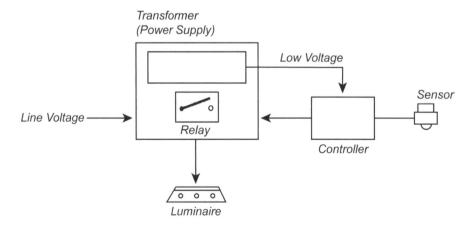

Figure 5-28   (EL9) Occupancy-Sensing Control

ASHRAE/IES Standard 90.1-2010 Occupancy Sensor Requirements

Occupancy sensors are required in ASHRAE/IES Standard 90.1-2010 in the following spaces: classrooms; conference, meeting, and training rooms; employee lunch and break rooms; storage and supply rooms between 50 and 1000 ft^2; rooms used for document copying and printing; office spaces up to 250 ft^2; restrooms; and dressing, locker, and fitting rooms.

Standard 90.1-2010 requires that occupancy sensors either shall be manual ON or shall be controlled to automatically turn the lighting on to not more than 50% power, except in public corridors, stairwells, restrooms, primary building entrance areas and lobbies, and areas where manual ON operation would endanger the safety or security of the room or occupants where full automatic ON is allowed.

Automatic Light Levels with Occupancy Sensors

Dual-Circuit Occupancy Sensor
Source: Michael Lane, Lighting Design Lab

To achieve the automatic ON to 50% light level, use a dual-circuit occupancy sensor with switch 1 set to automatic ON and switch 2 set to manual ON.

Circuit switch 1 to do one of the following:
* turn on half of the fixtures with half of the lamps in each fixture
* turn on all of the lamps in all of the fixtures to a 50% dim level

Circuit switch 2 to turn on the remainder of the lighting in the space.

In a small office with two fixtures, circuit switch 1 to control the fixture farthest from the window and switch 2 to control the fixture closest to the window.

ASHRAE/IES Standard 90.1-2010 Lighting Control Requirements

ASHRAE/IES Standard 90.1-2010 requires that in all spaces the controlled lighting shall have at least one control step between 30% and 70% (inclusive) of full lighting power in addition to all OFF except in corridors, electrical/mechanical rooms, public lobbies, restrooms, stairways, and storage rooms.

manual ON after either manual or automatic turn-off. Automatic time scheduling is an alternative to occupancy sensors in open-plan offices.

Private Offices, Conference Rooms, and Break Rooms. In private offices, conference rooms, and break rooms in daylight zones, dual-circuit infrared wall box sensors with integrated daylight override should be preset for manual ON/automatic OFF operation or automatic ON to 50%. In private offices, conference rooms, and break rooms not in daylight zones, use manual ON/automatic OFF without daylight override.

Other Areas. In nondaylighted areas, ceiling-mounted occupancy sensors are preferred.

EL10   *Multilevel Switching* (Climate Zones: all)

Specify luminaires with multiple lamps to be factory wired for inboard-outboard switching or in-line switching. The objective is to have multiple levels of light uniformly distributed in the space. Avoid checkerboard patterns of turning every other fixture off in medium and large spaces. In open office and large open areas, avoid nonuniform switching patterns unless different areas of the large space are used at different times or for different functions.

EL11   *Daylight-Responsive Controls* (Climate Zones: all)

Daylight controls are used in all open office areas in both the primary and secondary daylight zones. Locate a separate photocell in the center of each of the primary and secondary zones. Factory-setting of calibrations should be specified when feasible to avoid field labor. Lighting calibration and commissioning should be performed after furniture installation but prior to occupancy to ensure user acceptance.

EL12   *Exit Signs* (Climate Zones: all)

Use light-emitting diode (LED) exit signs or other sources that consume no more than 5 W per face. The selected exit sign and source should provide the proper luminance to meet all building and fire code requirements.

EL13   *Light Fixture Distribution* (Climate Zones: all)

Recessed high-performance lensed fluorescent fixtures should be used as the standard light fixture throughout the design. The high-performance lensed fixture is not the old-style 1960s flat prismatic lensed fluorescent fixture. Nor is it the "indirect basket" style fixture introduced by manufacturers in the late 1990s. These high-performance lensed fixtures were all introduced after 2005 and have fixture efficiencies of 85% or higher. Other fixture types, such as pendant-mounted direct/indirect, pendant-mounted direct, recessed parabolic, recessed indirect basket, and lensed fluorescent fixtures, may be able to meet the watts-per-square-foot allowance in the recommendations in Chapter 4, but the illuminance levels will be lower.

Sample Design Layouts for Office Buildings

The 0.75 W/ft^2 goal for LPD (shown in each recommendation table in Chapter 4) represents an average LPD for the entire building. Individual spaces may have higher power densities if they are offset by lower power densities in other areas. The example designs described below offer *a way*, but *not the only way*, that this watts-per-square-foot limit can be met. Daylighting (see DL17) is assumed in all open office plans and under all skylights.

The examples in EL14 through EL20 are based on a national "average" building space distribution. No building is average and each building will have a different space allocation. When following the recommendations below, adjust the standard space allocation to match the specific building's space allocation.

On a national average, a typical office building will have the following space distribution:

- Open office—16%
- Private offices—25%
- Conference rooms—10%
- Corridors—10%

Recessed High-Performance Lensed Fluorescent Fixtures

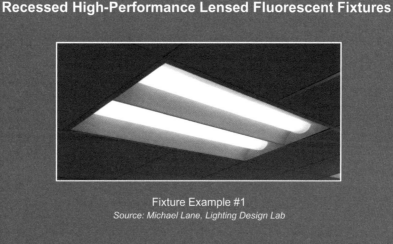

Fixture Example #1
Source: Michael Lane, Lighting Design Lab

Fixture Example #2
Source: Michael Lane, Lighting Design Lab

Recessed high-performance lensed fluorescent fixtures are available from more than six lighting manufacturers. The distinguishing feature of these fixtures is an optical lens (or lenses) that direct the light to the workstation. This lens is *not* the flat prismatic lens common to the old lensed fluorescent troffer.

- Storage— 15%
- Restrooms—4%
- Staff break rooms—3%
- Electrical/mechanical rooms—3%
- Stairways—2%
- Lobby—6%
- Other—6%

EL14 *Open-Plan Offices* (Climate Zones: all)

The target lighting in open offices is 30 average maintained footcandles for ambient lighting, with approximately 50 fc provided on the desktop by a combination of the ambient light and daylight. Supplemental task lighting is only required during nondaylight hours.

Open-plan office areas account for approximately 16% of the floor area and are designed to 0.68 W/ft^2, not including task lighting wattage (see EL5 for recommendations on task lighting). Assuming an 8 × 8 ft workstation and a layout that is only two deep from the window wall keeps all open office work areas inside the daylight zones. Use daylight dimming ballasts and separate photocell control in the primary and secondary daylight zones (see the definitions of primary and secondary daylight zones near DL3). Use occupancy sensor local control or scheduling on all luminaires (see EL9 and EL10). Depending on the height of the partitions that separate the corridor from the workstations, it is possible that the corridor will be lighted by the workstation light fixtures. Use compact fluorescent downlights or wall sconces to add additional illumination to the corridor as needed. See Figure 5-29 for an example open-plan office layout.

EL15   *Private Offices* (Climate Zones: all)

The target lighting in private offices is 30 average maintained footcandles for ambient lighting, with approximately 50 fc provided on the desktop by a combination of the ambient light and daylight. Supplemental task lighting is only required during nondaylight hours. A sample layout for a private office is shown in Figure 5-30. Use manual ON or automatic ON to 50% occupancy sensor local control.

Private offices account for approximately 29% of the floor area and are designed to 0.8 W/ft^2, not including task lighting wattage (see EL5 for recommendations on task lighting).

EL16   *Conference Rooms/Meeting Rooms* (Climate Zones: all)

The target lighting in conference rooms and meeting rooms is 30–40 average maintained footcandles. Use manual ON or automatic ON to 50% occupancy sensor local control.

Conference rooms account for approximately 10% of the floor area. The layout shown in Figure 5-31 is about 0.77 W/ft^2.

EL17   *Corridors* (Climate Zones: all)

The target lighting in corridors is 5–10 average maintained footcandles. Choose luminaires that light the walls and provide relatively uniform illumination.

Corridors account for approximately 10% of the floor area. Optional layouts using one-lamp 1 × 4 or 26 W compact fluorescent sconce or ceiling luminaires may be used to minimize the number of lamp types on the project. This layout yields 0.50 W/ft^2 when spaced 12 ft on center in a 5 ft wide corridor. Figure 5-32 shows a sample layout for corridors.

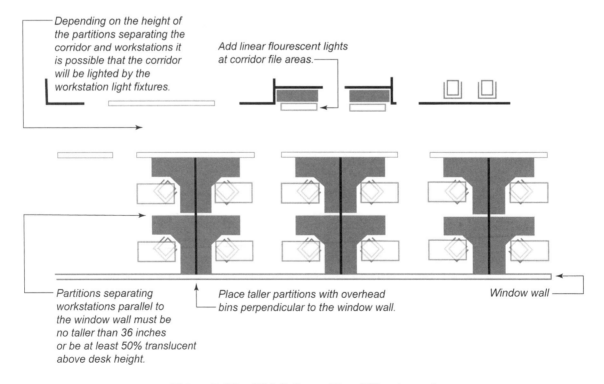

Figure 5-29   (EL14) Open-Plan Office Layout

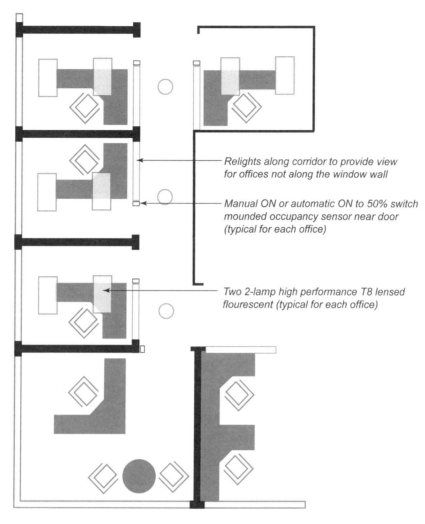

Relights along corridor to provide view for offices not along the window wall

Manual ON or automatic ON to 50% switch mounded occupancy sensor near door (typical for each office)

Two 2-lamp high performance T8 lensed flourescent (typical for each office)

Figure 5-30   (EL15) Private Office Layout

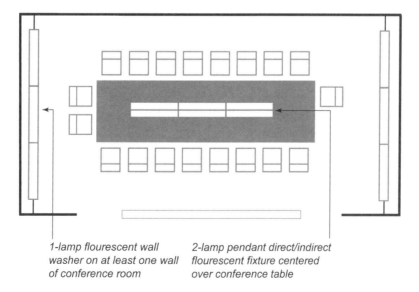

1-lamp flourescent wall washer on at least one wall of conference room

2-lamp pendant direct/indirect flourescent fixture centered over conference table

Figure 5-31   (EL16) Conference Rooms/Meeting Rooms Layout

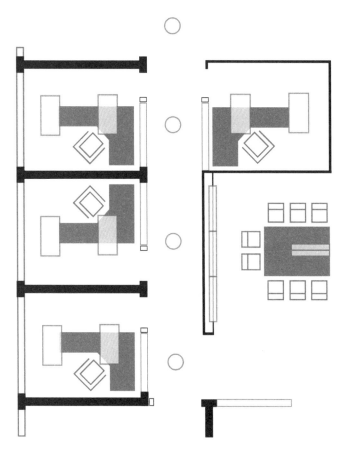

Figure 5-32   (EL17) Corridor Layout

EL18   *Storage Areas* (Climate Zones: all)

The target lighting in storage areas is 5–15 average maintained footcandles. Storage areas account for approximately 15% of the floor area and are designed to 0.64 W/ft^2. Figure 5-33 shows a sample layout for storage areas.

EL19   *Lobbies* (Climate Zones: all)

The target lighting in lobby areas is 10–15 average maintained footcandles. Highlight wall surfaces and building directories.

Lobbies account for approximately 6% of the floor area. The layout shown in Figure 5-34 is about 1.09 W/ft^2.

Note: Lighting in the remaining 10% of the office space is composed of various functions including restrooms, electrical/mechanical rooms, stairways, workshops, and others. Average the connected load in these spaces to 0.64 W/ft^2, which is equivalent to about one two-lamp high-performance T8 luminaire every 80 ft^2. Use occupancy sensors or timers where appropriate.

The designed LPD does not exceed the recommended 0.75 W/ft^2 for the total building interior.

EL20   *Twenty-Four-Hour Lighting* (Climate Zones: all)

Night lighting or lighting left on 24 hours to provide emergency egress needs when the building is unoccupied should be designed to limit the total lighting power to 10% of the total LPD (0.075 W/ft^2). It should be noted that most jurisdictions also allow the application of

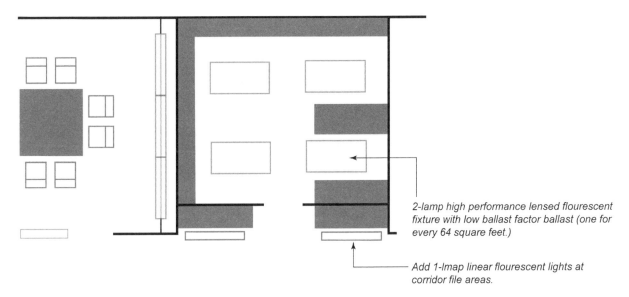

2-lamp high performance lensed flourescent fixture with low ballast factor ballast (one for every 64 square feet.)

Add 1-lmap linear flourescent lights at corridor file areas.

Figure 5-33 (EL18) Storage Area Layout

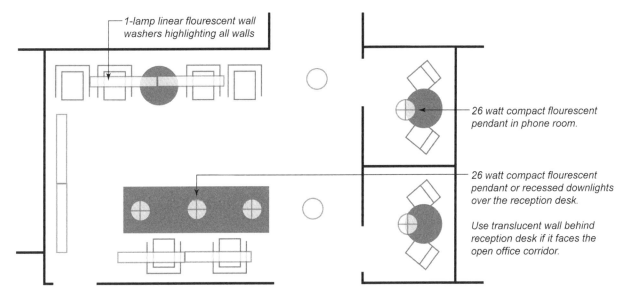

1-lamp linear flourescent wall washers highlighting all walls

26 watt compact flourescent pendant in phone room.

26 watt compact flourescent pendant or recessed downlights over the reception desk.

Use translucent wall behind reception desk if it faces the open office corridor.

Figure 5-34 (EL19) Lobby Layout

occupancy sensor controls on egress lighting to further reduce electricity associated with lighting an unoccupied building.

EXTERIOR LIGHTING

Good Design Practice

With the publication of ASHRAE/IES Standard 90.1-2010, exterior LPDs are now calculated using lighting zones (LZs). There are five zones, as shown in Table 5-5. These zones range from zone 0, which is "Undeveloped areas within national parks, state parks, forest land, rural areas, and other undeveloped areas," to zone 4, which is "High activity commercial districts in major metropolitan areas as designated by the local jurisdiction." Most office buildings will fall into zone 3.

Cautions: Calculate LPD only for areas intended to be lighted. For this Guide, areas that are lighted to less than 0.1 fc are assumed to not be lighted and cannot be counted in the LPD allowance. For areas that are intended to be lighted, design with a maximum to minimum ratio of illuminance no greater than 30 to 1. Therefore, if the minimum light level is 0.1, then the maximum level in that area should be no greater than 3 fc.

For parking lot and grounds lighting, do not increase luminaire wattage in order to use fewer lights and poles. Increased contrast makes it harder to see at night beyond the immediate fixture location. Flood lights and wall-packs should not be used, as they cause hazardous glare and unwanted light encroachment on neighboring properties.

Limit poles to 20 ft mounting height and use luminaires that provide all light below the horizontal plane to help eliminate light trespass and light pollution.

EL21 **Exterior Lighting Power—Parking Lots and Drives (Climate Zones: all)**

Limit exterior lighting power to 0.10 W/ft^2 for parking lots and drives in LZ3 and LZ4 or to 0.06 W/ft^2 in LZ2. Calculate only for paved areas, excluding grounds that are lighted to less than 0.1 fc.

Use LED parking lot fixtures with a bi-level switching driver that will reduce power between 12:00 p.m. and 6:00 a.m. to no more than 50%.

Cautions: Parking lot lighting locations should be coordinated with landscape plantings so that tree growth does not block effective lighting from pole-mounted luminaires.

Parking lot lighting should not be significantly brighter than the lighting of the adjacent street. Follow IES RP-33-1999 (IES 1999) recommendations for uniformity and illuminance recommendations.

EL22 **Exterior Lighting Power—Walkways (Climate Zones: all)**

Limit exterior lighting power to 0.08 W/linear foot for walkways less than 10 ft wide, to 0.16 W/ft^2 for walkways 10 ft wide or greater and plaza areas and special feature areas in LZ3 and LZ4, to 0.07 W/linear foot for walkways less than 10 ft wide, and to 0.14 W/ft^2 for walkways 10 ft wide or greater and plaza areas and special feature areas in LZ2. Exclude grounds that are lighted to less than 0.1 fc.

EL23 **Decorative Façade Lighting (Climate Zones: all)**

Avoid the use of decorative façade lighting. If façade lighting is desired, limit the lighting power to 0.075 W/ft^2 in LZ3 and LZ4 and to 0.05 W/ft^2 in LZ2 for the area intended to be illuminated to a light level no less than 0.1 fc.

Façade lighting that is installed is assumed to be programmed to turn off between the hours of midnight and 6:00 a.m. This does not include lighting of walkways or entry areas of the building that may also light the building itself.

Table 5-5   Exterior Lighting Zones

Lighting Zone	Description
0	Undeveloped areas within national parks, state parks, forest lands, rural areas, and other undeveloped areas as defined by the authority having jurisdiction
1	Developed areas of national parks, state parks, forest lands, and rural areas
2	Areas predominantly consisting of residential zoning, neighborhood business districts, light industrial buildings with limited nighttime use, and residential mixed-use areas
3	All other areas
4	High-activity commercial districts in major metropolitan areas as designated by the local jurisdiction

Source: ASHRAE (2010)

EL24 **Sources (Climate Zones: all)**

- All parking lot fixtures should use LED light sources.
- All grounds and building lighting should use pulse start metal halide, fluorescent, or compact fluorescent amalgam lamps with electronic ballasts.
- Standard high-pressure sodium lamps are not recommended due to their reduced visibility and poor color-rendering characteristics.
- Incandescent lamps are only recommended when used on occupancy sensors for lights that are normally off.
- For colder climates, fluorescent and compact fluorescent luminaires must be specified with cold-temperature ballasts. Use compact fluorescent amalgam lamps.

Exterior Lighting

Parking Lot Lighting Example
Source: Michael Lane, Lighting Design Lab

Parking lot lighting locations should be coordinated with landscape plantings so that tree growth does not block lighting from pole-mounted luminaires. In this example, note the separation of the lighting poles to the left of the walkway and the planting areas to the right of the walkway. Additionally, the trees are planted centered between the pole locations so when the trees are mature they will not block the light from lighting the parking lot.

The pole height is approximately 20 ft, and the luminaires are flat glass lenses that produce all light below the horizontal plane. These styles of luminaires help reduce light trespass and light pollution.

EL25   *Controls* (Climate Zones: all)

Use photocell or astronomical time switches on all exterior lighting. If a building energy management system is being used to control and monitor mechanical and electrical energy use, it can also be used to schedule and manage outdoor lighting energy use.

Turn off exterior lighting not designated for security purposes when the building is unoccupied by incorporating a time clock control. Design the total exterior lighting power (for parking, façades, building grounds, entry lights) to be reduced to 25% of the design level when no occupants are present between 12:00 p.m. and 6:00 a.m.

REFERENCES AND RESOURCES

ASHRAE. 2010. ANSI/ASHRAE/IES Standard 90.1-2010, *Energy Standard for Buildings Except Low-Rise Residential Buildings*. Atlanta: American Society of Heating, Refrigerating and Air-Conditioning.

DiLouie, C. 2007. *Lighting Controls Handbook*. NY: Illuminating Engineering Society of North America.

IES. 1994. IES DG-5-94, *Recommended Lighting for Walkways and Class 1 Bikeways*. NY: Illuminating Engineering Society of North America.

IES. 1998. IES RP-20-98, *Lighting for Parking Facilities*. NY: Illuminating Engineering Society of North America.

IES. 1999. IES RP-33-99, *Lighting for Exterior Environments*. NY: Illuminating Engineering Society of North America.

IES. 2003. IES G-1-03, *Guideline on Security Lighting for People, Property, and Public Spaces*. NY: Illuminating Engineering Society of North America.

IES. 2004. ANSI/IES RP-1-04, *American National Standard Practice for Office Lighting*. NY: Illuminating Engineering Society of North America.

IES. 2011. *The Lighting Handbook*, 10th ed. NY: Illuminating Engineering Society of North America.

NBI. 2011. Advanced Lighting Guidelines. www.algonline.org. White Salmon, WA: New Buildings Institute.

PLUG LOADS

EQUIPMENT AND CONTROL GUIDELINES

Good Design Practices

PL1   *Connected Wattage* (**Climate Zones: all**)

To reduce the connected wattage for plug load equipment, select office equipment and appliances that achieve low energy usage. ENERGY STAR®-rated equipment typically has significantly lowered operational wattage and may include improved sleep mode algorithms (EPA 2009). Desktop computers, laptops, desktop printers, fax machines, copy machines, refrigerators, microwave ovens, coffeemakers, and dishwashers are typical equipment types used in offices that have ENERGY STAR ratings. Look for efficient equipment even if not rated by ENERGY STAR, such as high-output photocopying machines.

PL2   *Laptop Computers* (**Climate Zones: all**)

Laptop computers are designed to operate efficiently to extend battery life. Efficient operation includes lower connected wattage and effective power management. Laptops with ENERGY STAR ratings should be selected. Desktop computers generally use significantly more energy and may not be necessary for most users.

Application of PL1 and PL2 can result in substantial reduction in equipment power density relative to a representative building with non-ENERGY STAR and other unrated equipment with half of the computers desktops and half laptops. The equipment power density values shown in Table 5-6 for a simulated medium office building of 53,600 ft^2 show a reduction from 0.75 W/ft^2 to 0.55 W/ft^2 based on total wattage of equipment for the entire building, not just actual office area (Roberson et al. 2004).

Table 5-6   Reduction in Equipment Wattage for PL1 and PL2

Plug Load Equipment Inventory	Baseline			Advanced		
	Qty	Plug Load, Each, W	Plug Load, W	Qty	Plug Load, Each, W	Plug Load, W
Computer—servers	8	65	520	8	54	432
Computer—desktop[a]	134	65	8710	89	54	4806
Computer—laptop[a]	134	19	2546	179	17	3043
Monitor—server—LCDs	8	35	280	8	24	192
Monitor—desktop—LCDs	268	35	9380	268	24	6432
Laser printer—network	8	215	1720	8	180	1440
Copy machine	4	1100	4400	4	500	2000
Fax machine	8	35	280	8	17	136
Water cooler	8	350	2800	8	193	1544
Refrigerator	8	76	608	8	65	520
Vending machine	4	770	3080	4	770	3080
Coffeemaker	4	1050	4200	4	1050	4200
Portable heaters, fans, etc.[b]	30	30	900	30	30	900
Other small appliances, chargers	250	4	1000	250	4	1000
Total plug load, W	40,424			29,725		
Plug load density, W/ft^2	0.75			0.55		

[a] Assumes shift toward higher proportion of laptops instead of desktop computers.
[b] Values shown per unit are not full connected wattage and are reduced for estimates of seasonal usage and averages for a mix of devices.

PL3   *Occupancy Controls* (Climate Zones: all)

Consider the use of occupancy-based control of devices to reduce energy consumption when the equipment is not in use. Occupancy-sensor-controlled plug strips can be used to power down monitors and other items typically plugged in at individual workstations, such as fans, chargers, and task lighting (task lighting control is described in detail in EL5). Another approach well suited to individual enclosed offices is to control selected outlets with a room-based occupancy sensor. This approach can also reduce parasitic losses—small amounts of electricity used by appliances even when the appliances are switched off. Good education can encourage occupants to plug the majority of their appliances into the occupancy-controlled plugs.

Vending machines should be equipped with occupancy sensor control for lighting and for cooling operation, if applicable. ENERGY STAR-rated vending machines include this type of control or can be retrofitted with add-on equipment.

Timer switches should be applied for central equipment that is unused during unoccupied periods but that should be available throughout occupied periods. Example equipment includes water coolers and central coffeemakers.

Computer power management allows computers to go into minimum energy usage when not active or turn off during scheduled hours. Individual devices with low power sleep modes can be purchased, such as ENERGY STAR-rated laptops. Power management can be fully activated in devices that may not use these modes in their default setup. Network power management software can allow central control for scheduled off hours and full activation of available power-saving modes while allowing the network management to turn units on for computer updates and maintenance.

PL4   *Parasitic Loads* (Climate Zones: all)

Reduce and eliminate parasitic loads. These loads include small energy usage from equipment that is nominally turned off but still using a trickle of energy. Transformers that provide some electronic devices with low-voltage direct current from alternating current plugs also draw power even when the equipment is off. Transformers are available that are more efficient and have reduced standby losses. Wall switch control of power strips, noted in PL3, cuts off all power to the plug strip, eliminating parasitic loads at that plug strip when the switch is controlled off. Newer power management surge protector outlet devices have low or no parasitic losses (Lobato et al. 2011).

PL5   *Printing Equipment* (Climate Zones: all)

Consolidate printing services to minimize the number of required devices. One study provides information on a strategy being applied in a new office building to eliminate personal printing devices and consolidate printing services to a smaller number of multifunction devices that reduces connected power and provides anticipated reduction in energy usage (Lobato et al. 2011). Use of multifunction devices that provide printing, copying, and faxing capabilities reduces power demand from multiple devices. The study indicates that printing services from these devices will be provided with 1 device per 60 occupants rather than 1 device per 40 occupants, as in the building occupants' old space.

PL6   *Unnecessary Equipment* (Climate Zones: all)

Determine whether there are pieces of equipment that operate as "nice-to-have" but which are not fundamental to the core function of the business. For example, large flat-screen TV arrays in lobby areas can be eliminated, and mechanically cooled drinking water can be replaced with filtered tap water.

REFERENCES AND RESOURCES

EPA. 2009. *ENERGY STAR Qualified Products.* www.energystar.gov/index.cfm?fuseaction=find_a_product.html. Washington, DC: U.S. Environmental Protection Agency.

Hart, R., S. Mangan, and W. Price. 2004. Who left the lights on? Typical load profiles in the 21st century. 2004 ACEEE Summer Study on Energy Efficiency in Buildings. August 22–27, Pacific Grove, CA.

Lobato, C., S. Pless, M. Sheppy, and P. Torcellini. 2011. Reducing plug and process loads for a large-scale, low-energy office building: NREL's Research Support Facility. *ASHRAE Transactions* 117(1):330–39.

Maniccia, D., and A. Tweed. 2000. *Occupancy Sensor Simulations and Energy Analysis for Commercial Buildings*. Troy, NY: Lighting Research Center, Renssaelaer Polytechnic Institute.

Roberson, J.A., C. Webber, M. McWhinney, R. Brown, M. Pinckard, and J. Busch. 2004. After-hours power status of office equipment and energy use of miscellaneous plug-load equipment. LBNL-53729. Berkeley: Lawrence Berkeley National Laboratory.

Sanchez, M.C., C.A. Webber, R. Brown, J. Busch, M. Pinckard, and J. Roberson. 2007. Space heaters, computers, cell phone chargers: How plugged in are commercial buildings? LBNL-62397. *Proceedings of the 2006 ACEEE Summer Study on Energy Efficiency in Buildings*, August, Asilomar, CA.

Thornton, B.A., W. Wang, M.D. Lane, M.I. Rosenberg, and B. Liu. 2009. *Technical Support Document: 50% Energy Savings Design Technology Packages for Medium Office Buildings*, PNNL-19004. Richland, WA: Pacific Northwest National Laboratory. www.pnl.gov/main/publications/external/technical_reports/PNNL-19004.pdf.

Thornton, B.A., W. Wang, Y. Huang, M.D. Lane, and B. Liu. 2010. *Technical Support Document: 50% Energy Savings for Small Office Buildings,* PNNL-19341. Richland, WA: Pacific Northwest National Laboratory. www.pnl.gov/main/publications/external/technical_reports/PNNL-19341.pdf.

SERVICE WATER HEATING

GENERAL RECOMMENDATIONS

Good Design Practice

WH1   *Service Water Heating Types* (Climate Zones: all)

The service hot water requirements for office buildings are typically very low and confined to restrooms, janitor closets, and break rooms. The most difficult problem, typically, is the wait time for hot-water delivery, especially when the water heater is remote from the end use. The typical remedy for long wait times for service hot water systems is a pumped return to ensure immediate hot-water delivery. The typical service hot water load in an office building is so low, however, that pump return energy and heat loss through the piping may outweigh the actual energy consumption for producing the required hot water.

The service water heating (SWH) equipment included in this Guide are for gas-fired water heaters and electric water heaters. Natural gas and propane fuel sources are available options for gas-fired units.

Reference will be made to provisions for solar SWH or heat recovery, but detailed guidelines are not given.

WH2   *System Descriptions* (Climate Zones: as indicated below)

Gas-fired storage water heater (Climate Zones: all): a water heater with a vertical or horizontal water storage tank. A thermostat controls the delivery of gas to the heater's burner. The heater requires a vent to exhaust the combustion products. An electronic ignition is recommended to avoid the energy losses from a standing pilot.

Gas-fired instantaneous water heater (Climate Zones: all): a water heater with minimal water storage capacity. Such heaters require vents to exhaust the combustion products. An electronic ignition is recommended to avoid the energy losses from a standing pilot. Instantaneous, point-of-use water heaters should provide water at a constant temperature regardless of input water temperature.

Electric resistance storage water heater (Climate Zones: all): a water heater consisting of a vertical or horizontal storage tank with one or more immersion heating elements. Thermostats controlling heating elements may be of the immersion or surface-mounted type.

Electric resistance instantaneous water heater (Climate Zones: all): a compact under-cabinet or wall-mounted water heater with an insulated enclosure and minimal water storage capacity. A thermostat controls the heating element, which may be of the immersion or surface-mounted type. Instantaneous, point-of-use water heaters should provide water at a constant temperature regardless of input water temperature.

Heat pump electric water heater (Climate Zones: ❶ ❷ ❸): a storage-type water heat using rejected heat from a heat pump as the heat source. Water storage is required because the heat pump is typically not sized for the instantaneous peak demand for service hot water, even in an office building. The heat source for the heat pump may be the interior air (such as that from a motor or switching room), which is beneficial in cooling-predominant climates; the circulating loop for a water-source heat pump (WSHP) system, also beneficial in cooling dominated climates; or a ground-coupled hydronic loop. Heat pump water heaters should show a coefficient of performance (COP) of at least 3.0.

WH3   *Sizing* (Climate Zones: all)

The water heating system should be sized to meet the anticipated peak hot-water load, typically for office buildings about 0.4 gallons per person per hour. Calculate the hot-water demand based on the sum of the fixture units served by each unit according to local code.

While hot-water temperature requirements for restrooms and break rooms of an office building vary by local and state codes within the range of 100°F to 120°F, note that production of service hot water at temperatures below approximately 135°F may result in bacterial growth within storage-type water heaters.

WH4   *Equipment Efficiency* **(Climate Zones: all)**

Efficiency levels are provided in this Guide for gas instantaneous, gas-fired storage, and electric resistance storage water heaters. For gas-fired instantaneous water heaters, the energy factor (EF) and thermal efficiency (E_t) levels are the same because there are no standby losses. The incorporation of condensing technology is recommended for gas instantaneous water heaters to achieve a minimum E_t of 0.90% and an EF of 0.90.

The recommended efficiency levels for gas-fired storage water heaters also require condensing technology (E_t > 90% or EF > 0.80).

The construction of a condensing water heater as well as the water heater venting must be compatible with the acidic nature of the condensate for safety reasons. Disposal of the condensate should be done in a manner compatible with local building codes.

Efficiency metrics for high-efficiency electric storage water heaters (EFs) are also provided in this Guide. These efficiency metrics represent premium products that have reduced standby losses. Table 5-7 summarizes required EFs for electric storage water heaters of various storage capacities based upon the equation for EF shown in the recommendation tables in Chapter 4 .

Electric instantaneous water heaters are an acceptable alternative to high-efficiency storage water heaters. Electric instantaneous water heaters are more efficient than electric storage water heaters, and point-of-use versions minimize piping losses. However, their impact on building peak electric demand can be significant and should be taken into account during design. Where unusually high hot-water loads (e.g., showers) are present during periods of peak electrical use, electric storage water heaters are recommended over electric instantaneous water heaters.

WH5   *Location* **(Climate Zones: all)**

The water heater should be close to the hot-water fixtures to avoid the use of a hot-water return loop or of heat tracing on the hot-water supply piping. With gas-fired water heaters, flue and combustion air considerations and, in some jurisdictions, code requirements, may limit potential locations.

Accommodation of renewable or "free" heat sources will most often necessitate a centralized service hot water system, since these sources are likely to be from a single point. Similarly, full utilization of the renewable or "free" source often requires a storage tank since the heating load may not coincide with the heat available from the source. These systems should be designed carefully, so that parasitic energy consumption for circulation loops does not offset energy conservation gains from the renewable or "free" resource.

Table 5-7   Electric Water Heater Energy Factors

Storage Volume	EF Requirement
30 gal	0.95
40 gal	0.94
50 gal	0.93
65 gal	0.91
75 gal	0.90
80 gal	0.89
120 gal	0.85

WH6   ***Pipe Insulation*** **(Climate Zones: all)**

All SWH piping should be installed in accordance with accepted industry standards. Insulation levels should be in accordance with the recommendation levels in the climate-specific recommendation tables in Chapter 4, and the insulation should be protected from damage. Include a vapor retardant on the outside of the insulation.

RESOURCES

AHRI. 2010. GAMA Certification Programs. http://www.ahrinet.org/gama+certification+programs.aspx. Arlington, VA: Air-Conditioning, Heating, and Refrigeration Institute.

ASHRAE. 2007. *ASHRAE Handbook—HVAC Applications*. Atlanta: American Society of Heating, Refrigerating and Air-Conditioning Engineers.

ASHRAE. 2010. ANSI/ASHRAE/IES Standard 90.1-2010, *Energy Standard for Buildings Except Low-Rise Residential Buildings*. Atlanta: American Society of Heating, Refrigerating and Air-Conditioning Engineers.

HVAC SYSTEMS AND EQUIPMENT

HVAC SYSTEM TYPES

Although many types of HVAC systems could be used in office buildings, this Guide prescriptively covers the following five system types, each of which has demonstrated the ability to meet the 50% savings criteria through extensive computer simulations.

- Single-zone, packaged air-source heat pump systems with electric resistance supplemental heat and dedicated outdoor air systems (DOASs). Recall that this Guide uses *DOAS* and *100% outdoor air system* (*100% OAS*) interchangeably. (See HV3.)
- Single-zone, high-efficiency WSHPs or ground-source heat pumps (GSHPs) with DOASs. (See HV4 and HV5.)
- Multiple-zone, variable-air-volume (VAV) packaged direct-expansion (DX) rooftop units with either a hydronic heating system, including boiler, internal heating coil, and perimeter heating system (convectors or terminal heating coils); an indirect gas furnace; or an electric resistance internal heating source with perimeter electric convection heat and high-performance system controls. (See HV6.)
- Multiple-zone, VAV air-handling units with packaged air-cooled chiller and gas-fired boilers. (See HV7.)
- Fan-coils with an air-cooled chiller and a gas-fired boiler and a DOAS. (See HV8.)
- Radiant systems with DOASs. Chilled water (CHW) provided by an air-cooled chiller and hot water provided by a condensing boiler. (See HV9.)

Unique recommendations are included for each HVAC system type in the climate-specific recommendation tables in Chapter 4. It is noted that in some climate zones, achievement of the 50% savings criteria is dependent on higher-efficiency components.

Good Design Practice

HV1 *Cooling and Heating Loads* (Climate Zones: all)

Heating and cooling system design loads for the purpose of sizing systems and equipment should be calculated in accordance with generally accepted engineering standards and handbooks such as *ASHRAE Handbook—Fundamentals* (ASHRAE 2009a). Any safety factor applied should be done cautiously and applied only to a building's internal loads to prevent oversizing of equipment. If the unit is oversized and the cooling capacity reduction is limited, short cycling of compressors could occur and the system may not have the ability to dehumidify the building properly; in addition, oversized equipment may operate less efficiently.

Calculation of HVAC unit heating and cooling capacities should include both the capacity for meeting space loads and the capacity for heating, cooling, and dehumidifying the required maximum flow of ventilation air. Where energy recovery is used, the reduction of cooling and heating loads, and the subsequent reduction of mechanical equipment size, should be taken into account. For heating, the OA temperature must be heated to the room temperature and the heat required added to the building heat loss. On VAV systems, the minimum supply airflow to a zone must comply with local codes, ASHRAE Standard 62.1 (ASHRAE 2010a), and ASHRAE/IES Standard 90.1 (ASHRAE 2010b) and should be taken into account in calculating heating loads of the OA.

HV2 *Certification of HVAC Equipment* (Climate Zones: all)

Rating and certification by industry organizations is available for various types of HVAC equipment. In general, the certification is provided by industry-wide bodies that develop specific procedures to test the equipment to verify performance; ASHRAE/IES Standard 90.1 has requirements for units for which certification programs exist. Certifications that incorporate published testing procedures and transparency of results are much more reliable for predicting actual performance than are certifications that are less transparent. For types of equipment for which certification is available, selection of products that have been certified is highly recommended. For products for which certification is not available or that have not been subjected to certification available for their type of equipment, the products should be rigorously researched for backup for performance claims made by the supplier. The project team should determine by what procedure the performance data was developed and establish any limitations or differentials between the testing procedure and the actual use. Examples of equipment types that have recognized certifications include packaged heat pumps, packaged air-conditioning units, water chillers, gas furnaces and boilers, cooling towers, and water heaters.

HV3   *Single-Zone, Packaged Air-Source Heat Pump Systems (or Split Heat Pump Systems) with Electric Resistance Supplemental Heat and DOASs* (Climate Zones: all)

In this system, a separate packaged heat pump unit (or split heat pump fan-coil) is used for each thermal zone. This type of equipment is available in preestablished increments of capacity. The components are factory designed and assembled and may include outdoor-air and return-air dampers, fans, filters, a heating source, a cooling coil, a compressor, controls, and an air-cooled condenser. The heating source is provided by reversing the refrigeration circuit to operate the unit as a heat pump to be supplemented by electric resistance heating if heat pump heating capacity is reduced below required capacity by low exterior air temperatures. Indirect-fired gas furnaces can be used as an alternative heat source with heat pumps but cannot operate to supplement the heat pump output. This alternative was not evaluated for this Guide.

The systems evaluated for this Guide treat recirculated air only, so that the unit fans can be cycled with load without interrupting ventilation air supply. Ventilation air is provided by a DOAS. Heat pumps may be used with the DOAS (see HV10).

The components can be assembled as a single package (such as a rooftop unit) or a split system that separates the evaporator and condenser/compressor sections. Single packaged units are typically mounted on the roof or at grade level outdoors. Split systems generally have the indoor unit or units (including fan, filters, and coils) located indoors or in an unconditioned space and the condensing unit located outdoors on the roof or at grade level. The indoor unit may also be located outdoors; if so, it should be mounted on the roof curb over the roof penetration for ductwork to avoid installing ductwork outside the building envelope. The equipment should be located to meet the acoustical goals of the space while minimizing fan power, ducting, and wiring.

Performance characteristics vary among manufacturers, and the selected equipment should match the calculated heating and cooling loads (sensible and latent), also taking into account the importance of providing adequate dehumidification under part-load conditions (see HV11). The equipment should be listed as being in conformance with electrical and safety standards with its performance ratings certified by a nationally recognized certification program.

The fan energy is included in the calculation of the energy efficiency ratio (EER) for heat pump equipment, based upon standard rating procedures of Air-Conditioning, Heating, and Refrigeration Institute (AHRI) that include an assumed external air delivery pressure drop (AHRI 2007, 2008). Pressure drop in the air delivery system, including ductwork, diffusers, and grilles, should not exceed 0.7 in. w.c.

The cooling equipment, heating equipment, and fans should meet or exceed the efficiency levels listed in Table 5-8.

Of critical importance for this type of system is the minimum outdoor temperature at which the unit can provide the required heating capacity to meet the building heating load. Heat pump products are available that are rated to provide as much as 70% of their AHRI rated capacity (47°F outdoor dry-bulb temperature, 70°F indoor dry-bulb temperature) at –4°F OA temperature. In general, heat pump units selected for an application should be rated to provide

Table 5-8   Constant-Volume Heat Pump Efficiency Levels*

Primary Space Heating and Cooling		
Size Category, Air-Source Heat Pump	Cooling Efficiency	Heating Efficiency
<65,000 Btu/h	15.0 SEER 12.0 EER	9.0 HSPF
65,000 – 135,000 Btu/h	11.5 EER 12.8 IEER	47°F db/43°F wb outdoor air 3.4 COP 17°F db/15°F wb outdoor air 2.4 COP
135,000 – 240,000 Btu/h	11.5 EER 12.3 IEER	47°F db/43°F wb outdoor air 3.2 COP 17°F db/15°F wb outdoor air 2.1 COP
≥240,000 Btu/h	10.5EER 11.3 IEER	47°F db/43°F wb outdoor air 3.2 COP 17°F db/15°F wb outdoor air 2.1 COP

* SEER = seasonal energy efficiency ratio, EER = energy efficiency ratio, IEER = integrated energy efficiency ratio, HSPF = heating seasonal performance factor, db = dry bulb, wb = wet bulb, COP = coefficient of performance.

heating at the 99.6% heating design OA temperature for the site. The units should be sized to meet 100% of the building internal heating requirement (not including OA heating) at a 98% heating design OA temperature.

HV4   Water-Source Heat Pumps (WSHPs) (Climate Zones: all)

Typically, a separate WSHP is used for each thermal zone. This type of equipment is available in preestablished increments of capacity. The components are factory designed and assembled and include a filter, a fan, a refrigerant-to-air heat exchanger, a compressor, a refrigerant-to-water heat exchanger, and controls. The refrigeration cycle is reversible, allowing the same components to provide cooling or heating.

Individual WSHPs are typically mounted in the ceiling plenum over the corridor (or some other noncritical space) or in a closet next to the occupied space. The equipment should be located to meet the acoustical goals of the space and minimize fan power, ducting, and wiring. This may require that the WSHPs be located outside of the space.

Packaged WSHPs four tons and above should incorporate a two-stage or variable-speed compressor with variable-speed fans and a multistage thermostat. The unit should be controlled so that airflow will reduce with compressor staging.

The cooling, heating, and fan performance of the heat pump unit should meet or exceed the levels listed below.

Refrigeration:

- Two stage—lower stage at 70% and lower load, full-load greater than 70%
- Cooling EER, part load/full load 17.6/15.0, heating COP 5.7/5.0
- Single-stage with variable-speed motor, cooling EER 16.4, heating COP 5.2

Air Delivery: The fan energy is included in the calculation for the EER for WSHP equipment based on standard rating procedures (AHRI/ASHRAE 2005) that include an assumed external air delivery pressure drop. Pressure drop in the air delivery system, including ductwork, diffusers, and grilles, should not exceed 0.5 in. w.c.

Per ASHRAE/IES Standard 90.1 (ASHRAE 2010b), the WSHP unit should incorporate a solenoid valve to shut off flow of circulating loop water through the unit when the compressors are de-energized. The unit should also cycle fans when no conditioning is called for.

In a traditional WSHP system, all the heat pumps are connected to a common water loop. A closed-circuit cooling tower and a hot-water boiler are also installed in this loop to maintain the temperature of the water within a desired range. See HV14 for required equipment efficiencies. The circulation loop should have a variable-speed pump. The circulating loop may include a controller to reset circulating loop temperature according to exterior and operating conditions.

OA is conditioned and delivered by a separate dedicated ventilation system. This may involve ducting the OA directly to each heat pump, delivering it in close proximity to the heat pump intakes, or ducting it directly to the occupied spaces. Depending on the climate, the dedicated OA unit may include components to filter, cool, heat, dehumidify, or humidify the OA (see HV10).

HV5   *Ground-Coupled Water-Source Heat Pump (WSHP) System* (Climate Zones: all)

A variation of the WSHP system takes advantage of the high thermal capacitance of the earth to store heat rejected into the ground during the cooling system as a resource for winter heating. In general, successful implementation of a ground-coupled heat pump system requires relative balance between the amount of heat extracted from the ground for the heating cycle and the amount of heat rejected into the ground for the cooling cycle. An appropriately sized ground-coupling system will result in relatively lower heat rejection temperature during the summer compared with cooling tower heat rejection. While the lower heat extraction temperature of a ground-coupled system compared with fuel-fired makeup typically results in a lower COP, performance is improved by the fact that no energy is consumed for makeup heating.

Performance characteristics of selected GSHP units should be the same as the WSHP specifications listed in HV4. GSHP units used with a closed-loop well system should have a low-temperature heating capability allowing them to heat at an entering tempered water-loop temperature of no more than 30°F. The circulating fluid for a closed-loop ground-coupled heat pump system should incorporate an antifreeze additive to prevent icing of the loop.

External pressure drop for these units, as with the WSHPs, should be limited to 0.5 in. w.c. Following are some considerations for incorporation of a ground-coupled heat pump:

- Balance of winter heating loads with summer cooling loads.
- Accurate determination of heat diffusivity of earth in contact with the ground-coupled heat transfer system, at a minimum through use of a test well to determine the nature of ground strata and ground water levels.
- Adequate sizing of the ground-coupling system, using accurate ground thermal diffusivity information, to limit minimum supply water temperature during the winter and maximum supply water temperature during the summer.
- Appropriate design and control of the hydronic circulation system to optimize pumping energy and maximization of heat pump annual heating and cooling efficiency.

HV6   *Multiple-Zone, VAV Packaged DX Rooftop Units with a Hot-Water Coil,*
Indirect Gas Furnace, or Electric Resistance in the Rooftop Unit and
Convection Heat in the Spaces (Climate Zones: all)

In this system, a packaged DX rooftop unit serves several individually controlled zones. Each thermal zone has a VAV terminal unit that is controlled to maintain temperature in that zone. The components of the rooftop unit are factory designed and assembled and include OA and return-air dampers, filters, fans, a cooling coil, a heating source, compressors, a condenser, and controls. The components of the VAV terminal units are also factory designed and assembled and include an airflow-modulation device, controls, and possibly a heating coil, fan, or filter.

VAV terminal units are typically installed in the ceiling plenum above the occupied space or adjacent corridor. However, the equipment should be located to meet the acoustical goals of the space; fan power, ducting, and wiring should be minimized.

All the VAV terminal units served by each rooftop unit are connected to a common air distribution system (see HV20). Cooling is provided by the centralized rooftop unit. Heating is provided by an electric convection heater in each space.

The cooling equipment, heating equipment, and fans should meet or exceed the efficiency levels listed in Table 5-9 and in the recommendation tables in Chapter 4. The cooling equipment should also meet or exceed the part-load efficiency level, where shown.

Table 5-9 VAV DX Cooling-Only Equipment Efficiency Levels*

Size Category	Cooling Efficiency
<65,000 Btu/h	15.0 SEER
65,000 – 135,000 Btu/h	11.5 EER 12.8 IEER
135,000 – 240,000 Btu/h	11.5 EER 12.3 IEER
240,000 – 760,000 Btu/h	10.5 EER 11.3 IEER
≥760,000 Btu/h	9.7 EER 10.9 IEER

* SEER = seasonal energy efficiency ratio, EER = energy efficiency ratio, IEER = integrated energy efficiency ratio.

Indirect gas-fired furnaces providing morning warm-up should be condensing furnaces and have at least an 80% efficiency level as required in ASHRAE/IESNA Standard 90.1-2004 (ASHRAE 2004). Requirements for zone heating are shown in the recommendation tables in Chapter 4. For systems using gas-fired boilers, condensing boilers should be used.

For packaged VAV DX systems, fan power is usually incorporated into the EER calculation. To achieve the required level of energy efficiency, air supply and delivery systems for the packaged VAV units should be designed to require no more than 2.0 in. w.c. external static pressure (ESP) and should include variable-frequency drives (VFDs) or other features that result in improved part-load performance. A reduced design supply air temperature (SAT) of 50°F is used to lower system airflow and the resultant pressure drop. To enhance economizer operation, SAT is reset up to 58°F in climate zones 1–3 and 61°F in other climate zones.

Units will have air-side economizers in all climate zones except climate zone 1, with control based on either dry-bulb temperature sensors or enthalpy sensors. See the recommendation tables in Chapter 4 for the requirements in each climate zone.

For VAV systems, the minimum supply airflow to a zone must comply with local codes and the current versions of ASHRAE Standard 62.1 (for minimum OA flow) and ASHRAE/IES Standard 90.1 (for minimum turndown before reheat is activated) (ASHRAE 2010a, 2010b).

Use the recommendation tables in Chapter 4 to determine the required SAT control sequences and the requirements for indirect evaporative precooling or ventilation air heat recovery. Ventilation optimization, a combination of zone demand-controlled ventilation (DCV) and system ventilation reset using the provisions of ASHRAE Standard 62.1, reduces OA during operation.

HV7 *Multiple-Zone, VAV Air-Handling Units with Packaged Air-Cooled Chiller and Gas-Fired Boiler* (Climate Zones: all)

Requirements for this system are very similar to those of the packaged VAV DX system described in HV6, except that the cooling source, cooling distribution, and cooling to air heat exchange are separately specified. Efficiency of the chilled-water equipment should meet the requirements in HV14. Cooling coils should be selected for a minimum of 15° ΔT on the water side. Cooling coils should also be selected at no more than 450 ft/min air face velocity to minimize air pressure drop. The chilled-water distribution system should be designed to meet the requirements of HV29. In order to achieve the required level of energy efficiency for air supply, the diffusers, duct system, return air path, coils, and filters should be selected for minimum pressure drop. Air delivery system pressure drop, fan selection for mechanical efficiency, and motor selection for efficiency should result in no more than 0.72 W/cfm at design airflow. This is derived from

- 3.5 in. of total static pressure,
- 65% fan efficiency,
- 93% motor efficiency, and
- 95% variable-speed drive (VSD) efficiency.

HV8   *Fan-Coils* (Climate Zones: all)

In fan-coil systems, a separate fan-coil unit is used for each thermal zone. The components are factory designed and assembled and include filters, a fan, heating and cooling coils, controls, and possibly OA and return-air dampers. OA supply is provided by a DOAS.

Fan-coils are typically installed in each conditioned space, in the ceiling plenum above the corridor (or some other noncritical space), or in a closet adjacent to the space. However, the equipment should be located to meet the acoustical goals of the space; this may require that the fan-coils be located outside of the space while also attempting to minimize fan power, ducting, and wiring. Fan-coils should be equipped with a variable-speed fan to automatically enable VAV operation and enhance motor efficiency.

All the fan-coils are connected to a common water distribution system. Cooling is provided by a centralized water chiller. Heating is provided by either a centralized boiler or by electric resistance heat located inside each fan-coil.

OA is conditioned and delivered by a DOAS that may involve ducting the OA directly to each fan-coil or ducting it directly to the occupied spaces. Depending on the climate, the dedicated OA unit may include a heat recovery device. (See HV10 and HV12.)

The cooling equipment, heating equipment, and fans should meet or exceed the efficiency levels listed in the recommendation tables in Chapter 4 or listed in this chapter (HV14, HV23). The cooling equipment should also meet or exceed the part-load efficiency level, where shown. Following are performance requirements for ducted fan-coils:

* 0.30 W/cfm design supply air to space with VAV operation
* Coil ΔTs of at least 14°F

HV9   *Radiant Heating and Cooling and DOAS* (Climate Zones: all)

In this system, a high-efficiency chilled-water system distributes water to radiant cooling panels or to tubing imbedded in floor slabs in each thermal zone to provide local cooling. Ventilation air is provided by a DOAS. The energy efficiency of radiant heating and cooling systems derives from two characteristics of the system:

* The extensive surface area of the systems allows heating and cooling loads to be met with very low-temperature hot water and relatively high-temperature CHW. Chilled water is typically supplied to the system at a minimum temperature of 60°F, while heating loads can usually be met with a maximum hot-water temperature of 95°F.
* Heating or cooling energy is transferred to the space with no energy expenditure for moving air. Heat transfer is entirely by natural convection and radiant means.

Radiant heating and cooling systems may be implemented through the use of ceiling-mounted radiant panels that affix water tubing to a ceiling tile. This tubing is served by water piping above the ceiling. Valving controls water flow to sections of the ceiling to provide temperature control in the space. If the ceiling radiant system is used for both heating and cooling, the ceiling may be divided into interior and perimeter zones with four pipes (hot and chilled water) to the perimeter zones.

Another form of the radiant heating and cooling system uses polymer tubing imbedded in concrete floor slabs. This approach has been applied to radiant heating for a number of years but recently has also been extended to space cooling. The primary issue for floor-slab radiant heating and cooling is changeover between the two modes of operation. Control systems must be designed with a significant dead band so that rapid changeover (with time-lagged system "fighting" in the slab) can be avoided. Typically, radiant heating and cooling floor-slab systems should not be controlled directly by air thermostats, because transient conditions may result in frequent changeover. Well-designed systems take advantage of the thermal capacitance of the floor slab to mitigate transient loads and provide consistent interior comfort conditions.

Radiant ceilings are less effective for space heating than are radiant floors, while radiant floors are less effective for space cooling, except for offsetting cooling loads from direct solar gain on the cooling floor. Radiant ceilings are more often seen in office spaces, while radiant floors are seen in lobbies, atriums, and circulation spaces.

The radiant system, however, provides only sensible heating and cooling. All humidity control (both dehumidification and humidification) must be provided by a DOAS. In some cases, the required dew-point temperature of the incoming ventilation air may be lower than common practice in order to achieve the necessary dehumidification. For radiant cooling systems, especially in humid climates, avoidance of condensation on the cooling surfaces is the most important design consideration. Mechanisms for avoiding condensation include the following:

- Control of entering dew-point temperature of ventilation air to meet maximum interior air dew-point temperature limits.
- Design of radiant cooling systems to meet sensible cooling loads with elevated (>60°F) chilled-water temperatures.
- Monitoring of space dew-point temperature with radiant system shutdown upon detection of elevated space dew-point temperature.
- Design of building envelope systems to minimize infiltration. Construction-phase quality control of envelope systems to meet infiltration specifications.
- Removal of radiant cooling elements from areas immediately surrounding exterior doors.
- Provision of excess dehumidified ventilation air adjacent to likely sources of exterior air infiltration.

Passive chilled beams behave similarly to radiant chilled ceilings. The same issues of condensation avoidance and control apply. Passive chilled beams are also relatively ineffective for space heating.

Radiant Heating/Cooling System

Radiant Floor Tubing Layout
Source: WSP Flack + Kurtz

The atrium of the Syracuse University School of Management in Syracuse, New York, employs a radiant heating and cooling system. The tubing is covered by a concrete topping slab and then stone pavers. The system does only sensible heating and cooling; ventilation and dehumidification are provided through displacement diffusers located in the sidewalls of the large triangular planters in the space. The tubing is arrayed in the double serpentine pattern that is characteristic for radiant cooling slab applications.

Design of radiant systems that incorporate cooling is a very specialized task that should only be undertaken with experienced engineering input. Issues of temperature control, load response, condensation avoidance, etc., are likely to be peculiar to each project and may require custom solutions.

HV10 *Dedicated Outdoor Air Systems (100% Outdoor Air Systems)* (Climate Zones: all)

Dedicated outdoor air systems (DOASs) can reduce energy use by decoupling the dehumidification and conditioning of OA ventilation from sensible cooling and heating in the zone. The OA is conditioned by a separate DOAS that is designed to dehumidify the OA and to deliver it dry enough (with a low enough dew point) to offset space latent loads, thus providing space humidity control (Mumma 2001; Morris 2003). The DOAS also can be equipped with high-efficiency filtration systems with static pressure requirements above the capability of fan-coils. Terminal HVAC equipment heats or cools recirculated air to maintain space temperature. Terminal equipment may include fan-coil units, WSHPs, zone-level air handlers, or radiant heating and/or cooling panels.

There are many possible 100% outdoor air system (100% OAS) configurations (see Figure 5-35 for a few typical ones). The salient energy-saving features of DOASs are the separation of ventilation air conditioning from zone air conditioning and the ease of implementation of energy recovery.

DOASs can also be used in conjunction with multiple-zone recirculating systems, such as centralized VAV air handlers, but most often VAV systems do not use separate ventilation systems.

DOASs can reduce energy use in primarily three ways: 1) they often avoid the high OA intake airflows at central air handlers needed to satisfy the multiple spaces equation of ASHRAE Standard 62.1 (ASHRAE 2010a), 2) they eliminate (or nearly eliminate) simultaneous cooling and reheat that would otherwise be needed to provide adequate dehumidification in humid climates, and 3) with constant-volume zone units (heat pumps, fan-coils), they allow the unit to cycle with load without interrupting ventilation airflow. A drawback of many DOASs is that they cannot provide air-side economizing. This is more significant in drier climates where 100% OA can be used for economizing without the concern of raising indoor humidity levels. In addition, population diversity and use of "unused outdoor air" are not allowed when ventilation is provided separately.

Consider delivering the conditioned OA cold (not reheated to neutral) whenever possible and reheat only when needed. Providing cold (rather than neutral) air from the DOAS offsets a portion of the space sensible cooling loads, allowing the terminal HVAC equipment to be downsized and use less energy (Mumma and Shank 2001; Murphy 2006). Reheating the dehumidified air (to a temperature above the required dew point) may be warranted

- if the reheat consumes very little energy (using energy recovery, solar thermal source, etc.) and none of the zones are in the cooling mode,
- if all of the zones are in the heating mode, or
- if, for those zones in the cooling mode, the extra cooling energy needed (to offset the loss of cooling due to delivering neutral-temperature ventilation air) is offset by higher-efficiency cooling equipment and the reduction in heating energy needed for those zones in the heating mode (this is more likely to be true on an annual basis if the reheat in the DOAS is accomplished via air-to-air or condenser heat recovery).

In addition, implementing reset control strategies and exhaust air energy recovery (see HV12) can help minimize energy use.

For office buildings, dedicated OA units may be combined with terminal units such as fan-coils, radiant systems, or WSHPs that provide zone control. When used with a WSHP system, the DOAS can be served by a water-to-water heat pump system with performance as specified in the climate-specific recommendation tables in Chapter 4.

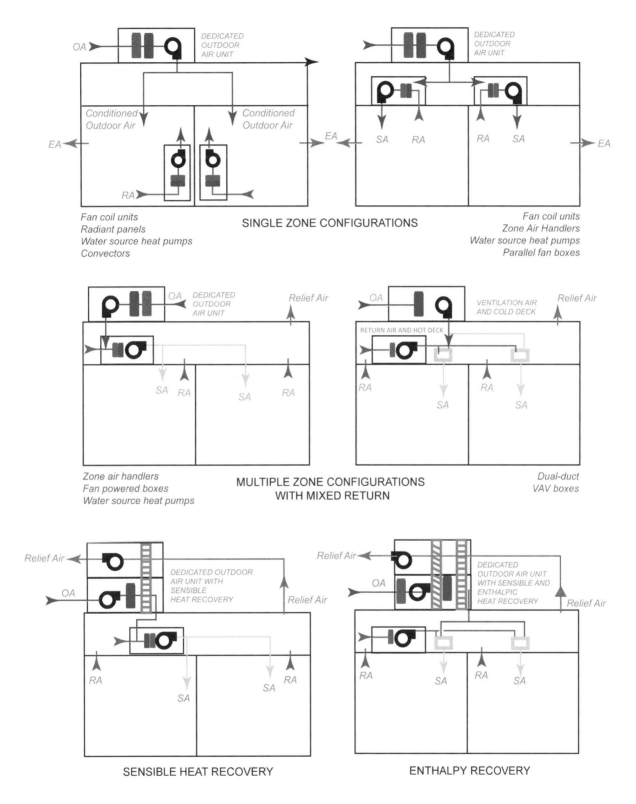

Figure 5-35 (HV10) Examples of DOAS Configurations

The cooling equipment, heating equipment, and fans should meet or exceed the efficiency levels listed in Table 5-10. The cooling equipment should also meet or exceed the part-load efficiency level, where shown. Presently there is no energy performance rating standard for 100% OASs. In order to meet the 50% energy savings of this Guide, all 100% OAS must incorporate energy recovery systems so that the incoming air condition is similar to the condition of incoming mixed air in a conventional system. This Guide specifies 100% OAS performance (with heat recovery) at the standard AHRI conditions used to determine EERs and heating seasonal performance factors (HSPFs) for conventional mixed-air systems. Gas furnaces should be condensing.

Exhaust-air heat recovery is necessary for use of heat pumps with DOASs in locations with low design heating temperatures. Most heat pump units do not operate well with low entering indoor air temperatures. The use of exhaust heat recovery will raise the incoming ventilation air to a temperature more compatible with heat pump operation.

Systems delivering 100% OA have many different configurations. In general, the air delivery system should be configured for no more than 1.5 in. w.c. total static pressure drop. For units that do not have EER ratings per AHRI, fans should be selected for a minimum 65% mechanical efficiency and motors at no less than 93% efficiency.

HVAC SYSTEM CONSIDERATIONS

HV11   Part-Load Dehumidification (Climate Zones: all)

Most basic, constant-volume systems (small packaged rooftop units, DX split systems, fan-coils, WSHPs, etc.) supply a zone with a constant amount of air regardless of the cooling load. The system must deliver warmer air under part-load conditions to avoid overcooling the

Table 5-10   DOAS Cooling and Heating Equipment Efficiencies*

Primary Space Heating and Cooling			
Size Category	Cooling Efficiency	Heating Efficiency	
Air-Source Heat Pump			
<65,000 Btu/h	15.0 SEER 12.0 EER	9.0 HSPF	
65,000 – 135,000 Btu/h	11.5 EER 12.8 IEER	47°F db/43°F wb outdoor air	3.4 COP
		17°F db/15°F wb outdoor air	2.4 COP
135,000 – 240,000 Btu/h	11.5 EER 12.3 IEER	47°F db/43°F wb outdoor air	3.2 COP
		17°F db/15°F wb outdoor air	2.1 COP
≥ 240,000 Btu/h	10.5 EER 11.3 IEER	47°F db/43°F wb outdoor air	3.2 COP
		17°F db/15°F wb outdoor air	2.1 COP
VAV DX, Gas Heat			
<65,000 Btu/h	15.0 SEER	80%	
65,000 – 135,000 Btu/h	11.5 EER 12.8 IEER	80%	
135,000 – 240,000 Btu/h	11.5 EER 12.3 IEER	80%	
240,000 – 760,000 Btu/h	10.5 EER 11.3 IEER	80%	
≥760,000 Btu/h	9.7 EER 10.9 IEER	80%	
Water-to-Water Heat Pump			
Any size	13.8 EER, 86°F EWT	3.8 COP, 68°F EWT	
		3.1 COP, 50°F EWT	

* SEER = seasonal energy efficiency ratio, EER = energy efficiency ratio, IEER = integrated energy efficiency ratio, HSPF = heating seasonal performance factor, db = dry bulb, wb = wet bulb, EWT = entering water temperature, COP = coefficient of performance.

space. In a typical chilled-water constant-volume air-handling unit application, a modulating valve reduces system capacity by throttling the water flow rate through the cooling coil. The warmer coil surface that results provides less sensible cooling (raising the supply air dry-bulb temperature), but it also removes less moisture from the passing airstream (raising the supply air dew point). Fan-coil units, on the other hand, often have two-position valves that provide either full cooling or no cooling.

In a typical single-compressor DX application or in a fan-coil unit with a two-position chilled-water valve, coil cooling periodically cycles off completely (through de-energizing the compressor or closing the chilled-water valve) to avoid overcooling. For a typical system, during the period that the coil is inactive, underhumidified OA is delivered to the space. While de-energizing the fan during the time that cooling is disabled would avoid loss of humidity control, it would also fail to deliver the required ventilation volume to the space. With a conventional constant-volume system, the choice is among loss of humidity control, failure to meet ventilation standards, or reheat to maintain space dry-bulb temperature.

Briefly stated, a basic constant-volume system matches sensible capacity to the sensible load; dehumidification capacity is coincidental. As the load diminishes, the system delivers ever warmer supply air or periodically delivers completely uncooled and completely undehumidified air. Dehumidification occurs at a much reduced rate either because the coil is only intermittently active or because the temperature of the air off the coil (and probably the supply air dew-point temperature) is raised. As a result, in humid outdoor conditions, the space relative humidity will tend to increase under part-load conditions, unless reheat is used.

Following are some (but not all) of the possible methods for improving part-load dehumidification.

- *For single-zone heat pump packaged units or split systems (see HV3).* Packaged rooftop units (or split systems) should use a DOAS (see HV10) to dehumidify the OA so that it is dry enough (has a low enough dew point) to offset the latent loads in the spaces. This helps avoid high indoor humidity levels without additional dehumidification enhancements in the local heat pump units.

- *For WSHPs or GSHPs (see HV4 and HV5).* The DOAS (see HV10) should be designed to dehumidify the OA so that it is dry enough (has a low enough dew point) to offset the latent loads in the spaces. This helps avoid high indoor humidity levels without additional dehumidification enhancements in the WSHP units. Alternatively, some WSHPs could be equipped with hot-gas reheat for direct control of space humidity.

- *For fan-coil units (see HV8).* The DOAS (see HV10) should be designed to dehumidify the OA so that it is dry enough (has a low enough dew point) to offset the latent loads in the spaces. This helps avoid high indoor humidity levels without additional dehumidification enhancements in the fan-coil units. Alternatively, fan-coils could be equipped with variable-speed fans for improved part-load dehumidification or with a reheat coil for direct control of space humidity. Consider using recovered heat when using reheat.

- *For multiple-zone, packaged VAV rooftop units (see HV6).* VAV systems typically dehumidify effectively over a wide range of indoor loads, as long as the VAV rooftop unit continues to provide cool, dry air at part-load conditions. One caveat: use caution when resetting the SAT upward during the cooling season. Warmer supply air means less dehumidification at the coil and higher humidity in the space. If SAT reset is used, include one or more zone humidity sensors to disable the reset if the relative humidity within the space exceeds 60%.

- *For multiple-zone, VAV air handlers (see HV7).* VAV systems typically dehumidify effectively over a wide range of indoor loads, as long as the VAV rooftop unit continues to provide cool, dry air at part-load conditions. One caveat: use caution when resetting the SAT or chilled-water (CHW) temperature upward during the cooling season. Warmer supply air (or water) means less dehumidification at the coil and higher humidity in the space. If SAT or CHW reset is used in a humid climate, include one or more zone humidity sensors to disable reset if the relative humidity within the space exceeds 60%.

HV12 ***Exhaust Air Energy Recovery* (Climate Zones: all)**

Exhaust air energy recovery can provide an energy-efficient means of reducing the latent and sensible OA cooling loads during peak summer conditions. It can also reduce the OA heating load in mixed and cold climates. HVAC systems that use exhaust air energy recovery should to be resized to account for the reduced OA heating and cooling loads (see ASHRAE [2008a]).

For some HVAC system types, the climate-specific recommendation tables in Chapter 4 recommend either exhaust air energy recovery or DCV. If the energy recovery option is selected, this device should have a total effectiveness as shown in Table 5-11. Note that in some climates energy recovery is not required.

The performance levels in Table 5-11 should be achieved with no more than 0.85 in. w.c. static pressure drop on the supply side and 0.65 in. w.c. static pressure drop on the exhaust side.

Sensible energy recovery devices transfer only sensible heat. Common examples include coil loops, fixed-plate heat exchangers, heat pipes, and sensible energy rotary heat exchangers (sensible energy wheels). Total energy recovery devices transfer not only sensible heat but also moisture (or latent heat)—that is, energy stored in water vapor in the airstream. Common examples include total energy rotary heat exchangers (also known as *total energy wheels* or *enthalpy wheels*) and fixed-membrane heat exchangers (see Figure 5-36). Energy recovery devices should be selected to avoid cross-contamination of the intake and exhaust airstreams. For rotary heat exchangers, avoidance of cross-contamination typically includes provision of a purge cycle in the wheel rotation and maintenance of the intake system pressure higher than the exhaust system pressure.

An exhaust-air energy recovery device can be packaged in a separate energy recovery ventilator (ERV) that conditions the OA before it enters the air-conditioning unit or the device can be integral to the air-conditioning unit.

For maximum benefit, the system should provide as close to balanced outdoor and exhaust airflows as is practical, taking into account the need for building pressurization and any exhaust that cannot be ducted back to the energy recovery device.

Exhaust for energy recovery may be taken from spaces requiring exhaust (using a central exhaust duct system for each unit) or directly from the return airstream (as with a unitary accessory or integrated unit). (See also HV19.)

Where an air-side economizer is used along with an ERV, add bypass dampers (or a separate OA path) to reduce the air-side pressure drop during economizer mode. In addition, the ERV should be turned off during economizer mode to avoid adding heat to the outdoor airstream. Where energy recovery is used without an air-side economizer, the ERV should be controlled to prevent the transfer of unwanted heat to the outdoor airstream during mild outdoor conditions.

In cold climates, follow the manufacturer's recommendations for frost prevention.

HV13 ***Indirect Evaporative Cooling* (Climate Zones: ②B ③B ④B ⑤B)**

In dry climates, incoming ventilation air can be precooled using indirect evaporative cooling. For this strategy, the incoming ventilation air (the primary airstream) is not humidified; instead, a separate stream of air (the secondary or heat rejection stream) is humidified, dropping its temperature, and is used as a heat sink to reduce the temperature of the incoming ventilation air.

Table 5-11   Total System Effectiveness with Energy Recovery

Condition	Effectiveness		
	Sensible	Latent	Total
Heating at 100% airflow	78	70	75
Heating at 75% airflow	83	77	82
Cooling at 100% airflow	80	71	75
Cooling at 75% airflow	84	78	82

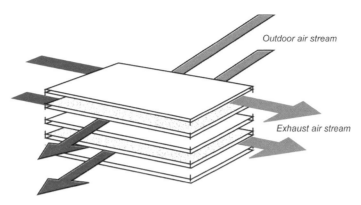

Fixed-plate heat exchanger (crossflow)

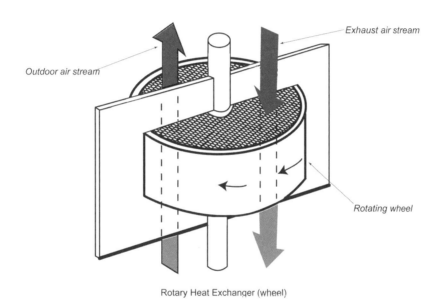

Rotary Heat Exchanger (wheel)

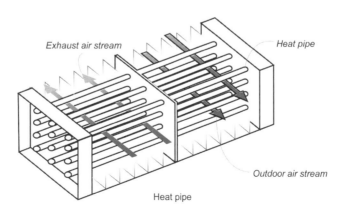

Figure 5-36   (HV12) Examples of Exhaust Air Energy Recovery Devices

The source of the heat rejection stream of air can be either OA or exhaust air from the building. If the air source is exhaust air, this system becomes an alternative for HV12.

Sensible heat transfer between the ventilation airstream and the evaporatively cooled secondary airstream can be accomplished using plate or tubular air-to-air heat exchangers, heat pipes, or a pumped loop between air coils in each stream (often called a *runaround loop*). For indirect evaporative coolers that use exhaust air as the secondary stream, the evaporative cooler can also function for sensible heat recovery during the heating season. If a runaround loop is used for heat transfer both for indirect evaporative cooling and heat recovery, the circulating fluid should incorporate antifreeze levels appropriate to the design heating temperature for that location.

Indirect evaporative cooling has the advantage that indoor air quality (IAQ) is not affected, as the evaporative cooling process is not in the indoor airstream. Air quality is not as critical for the exhausted secondary airstream as it is for the ventilation stream entering the occupied space.

Indirect evaporative coolers should be selected for at least 90% evaporative effectiveness for the evaporatively cooled airstream and for at least 65% heat transfer efficiency between the two airstreams.

Indirect evaporative coolers should also be selected to minimize air pressure drop through the heat exchangers

HV14   *Cooling and Heating Equipment Efficiencies* (Climate Zones: all)

The cooling and heating equipment should meet or exceed the efficiency levels listed in the climate-specific recommendation tables in Chapter 4. The cooling equipment should also meet or exceed the part-load efficiency level where shown. In some cases, recommended equipment efficiencies are based on system size (capacity).

There are many factors involved in making a decision whether to use gas or electricity, such as the availability of service, utility costs, operator familiarity, and the impact of source energy use. Efficiency recommendations for both types of equipment are provided in the recommendation tables in Chapter 4 to allow the user to choose.

Air-Cooled Chillers. Air-cooled water chillers should be certified or independently tested to produce a full-load EER of 10.0 or higher and an integrated part-load value (IPLV) of 12.5 or higher, according to AHRI rating methods. Chillers less than 40 tons should provide at least two steps of unloading while those 40 tons and above should provide at least four steps of unloading or continuous unloading. Chillers should incorporate controls capable of accommodating variable evaporator water flow while maintaining control of leaving chilled-water temperature. Water-cooled chillers and cooling towers were not analyzed for this Guide. A system including a water-cooled chiller, condenser water pump, and cooling tower all with sufficient efficiency and integrated controls may give the same or better energy performance as an air-cooled chiller.

Space-Heating Water Boilers. All gas-fired boilers specified for space heating should be of the condensing type with a minimum efficiency of 90% at 125°F return hot-water temperature. Zone heat transfer equipment should be sized based on 140°F entering hot-water temperature and as large a temperature drop through the air heating coil as possible. Boilers should be operated with a maximum leaving hot-water temperature of 140°F and should incorporate a leaving hot-water temperature reset control based on total heating load.

Closed-Circuit Cooling Towers (for WSHP Systems). Closed-circuit cooling towers should be selected for a maximum 10°F approach of cooling tower leaving water temperature to design wet-bulb temperature. Towers should also be selected for fan energy input at full load of no more than 120 W per nominal ton of capacity. Cooling tower fan motors larger than one horsepower should be equipped with VSDs.

HV15   *Ventilation Air* (Climate Zones: all)

The zone-level outdoor airflows and the system-level intake airflow should be determined based on the most recent edition of ASHRAE Standard 62.1 (ASHRE 2010a) but should not be less than the values required by local code unless approved by the authority having jurisdiction. The number of people used in computing the breathing zone ventilation rates should be based on known occupancy, local code, or the default values listed in Standard 62.1.

Cautions: The occupant load, or exit population, used for egress design to comply with the fire code is typically much higher than the zone population used for ventilation system design. Using occupant load rather than zone population to calculate ventilation requirements can result in significant overventilation, oversized HVAC equipment, and excess energy use.

Buildings with multiple-zone recirculating ventilation systems can be designed to account for recirculated OA as well as system population diversity using the Ventilation Rate Procedure of ASHRAE Standard 62.1 (ASHRAE 2007a). In effect, the multiple-zone recirculating ventilation system design approach allows ventilation air to be calculated on the basis of how many people are *in the building* (system population at design) rather than the sum of how many people are *in each space* (sum-of-peak zone population at design). Using the Ventilation Rate Procedure can reduce the energy required to condition ventilation air in office buildings. Refer to *62.1 User's Manual* for specific guidance (ASHRAE 2007b).

For all zones, time-of-day schedules in the building automation system (BAS) should be used to introduce ventilation air only when a zone is expected to be occupied.

For Radiant Systems (see HV9). Each zone served locally by the radiant system should be provided with outdoor ventilation air by a DOAS (see HV10).

For WSHPs or GSHPs (see HV4 and HV5). The DOAS (see HV10) should deliver the conditioned OA directly to each zone, to the intake of each individual heat pump (where it mixes with recirculated air, either in the ductwork prior to the heat pump or in a mixing plenum attached to the heat pump), or to the supply side of each WSHP (where it mixes with supply air from the heat pump before being delivered to the zone). Units should be configured such that ventilation air can be delivered to the space even when the unit fan is not running so that the unit can cycle for load without interrupting ventilation supply.

For Fan-Coil Units (see HV8). The DOAS (see HV10) should deliver the conditioned OA directly to each zone, to the intake of each individual fan-coil (where it mixes with recirculated air, either in the ductwork prior to the fan-coil or in a mixing plenum attached to the fan-coil), or to the supply side of each fan-coil (where it mixes with supply air from the fan-coil before being delivered to the zone). Fan-coils should be configured such that ventilation air can be delivered to the space even when the unit fan is not running so that the unit can cycle for load without interrupting ventilation supply.

For multiple-zone, packaged VAV rooftop units or VAV air handlers (see HV6 and HV7). Each rooftop unit should have an OA intake through which OA is introduced and mixes with the recirculated air prior to being delivered to the zones.

HV16   *Economizer* (Climate Zones: ❷ ❸ ❺ ❻ ❼ ❽)

Economizers, when recommended, help save energy by providing free cooling when ambient conditions are suitable to meet all or part of the cooling load. In humid climates, consider using enthalpy-based controls (versus dry-bulb temperature controls) to help ensure that unwanted moisture is not introduced into the space. See the climate-specific recommendation tables in Chapter 4 for economizer recommendations by climate zone.

Non-dedicated OA systems should be capable of modulating the OA, return air, and relief air dampers to provide up to 100% of the design supply air quantity as OA for cooling. (See HV10 for a discussion of DOASs.) A motorized OA damper should be used instead of a gravity damper to prevent unwanted OA from entering during unoccupied periods when the unit may recirculate air to maintain setback or setup temperatures. For all climate zones, the motorized OA damper should be closed during the entire unoccupied period, except when it may open in conjunction with unoccupied economizer cycle operation or a preoccupancy purge cycle.

Periodic maintenance is important with economizers, as dysfunctional economizers can cause substantial excess energy use because of malfunctioning dampers or sensors (see HV34).

HV17   *Demand-Controlled Ventilation (DCV)* **(Climate Zones: all)**

DCV can reduce the energy required to condition OA for ventilation. To maintain acceptable IAQ, the setpoints (limits) and control sequence must comply with ASHRAE Standard 62.1 (ASHRAE 2010a). Refer to Appendix A of *62.1 User's Manual* for specific guidance (ASHRAE 2007b).

For some HVAC system types, the climate-specific recommendation tables in Chapter 4 recommend either exhaust air energy recovery or DCV. If the DCV option is selected, the controls should vary the amount of OA in response to the need in a zone. The amount of OA could be controlled by 1) a time-of-day schedule in the BAS; 2) an occupancy sensor (such as a motion detector) that indicates when a zone is occupied or unoccupied; or 3) a carbon dioxide (CO_2) sensor, as a proxy for ventilation airflow per person, that measures the change in CO_2 levels in a zone relative to the levels in the OA. A controller will then operate the OA, return air, and relief air dampers to maintain proper ventilation. For options 1 and 2 above, ventilation rates for the occupied period should be based upon full occupancy and should be calculated in accordance with Section 6 of Standard 62.1-2007. For option 3 above, the full-load ventilation rate should be calculated according to Chapter 6 of Standard 62.1 and ventilation rate reductions should be controlled according to Informative Appendix C of Standard 62.1 (ASHRAE 2007a).

CO_2 sensors should be used in zones that are densely occupied and have highly variable occupancy patterns during the occupied period, such as conference rooms or meeting areas. For the other zones, occupancy sensors should be used to reduce ventilation when a zone is temporarily unoccupied. For all zones, time-of-day schedules in the BAS should be used to introduce ventilation air only when a zone is expected to be occupied.

Multiple-zone recirculating systems (such as VAV systems) require special attention to ensure adequate OA is supplied to all zones under varying loads. Employing DCV in a DOAS requires an automatic damper, a CO_2 sensor, and an airflow measurement device for each DCV zone. If the automatic damper selected is of the pressure-independent type, by definition it has flow measurement capability.

Control of a DCV system to match airflow volume to occupancy requires a continuous search algorithm for the controller. The controller continuously calculates the volume required in the space based upon the measured CO_2 concentration differential and then updates the volume setpoint to the pressure-independent damper controller. The following equation gives the required volume in the space based upon the measured CO_2 differential:

$$V'_{ot} = R_a \cdot A_z/(E_z - ((R_p \cdot (C_R - C_{OA}))/(8400 \cdot m))$$

where

V'_{ot} = required airflow volume at any point in time

R_a = zone area ventilation rate (0.06 cfm/ft^2 per ASHRAE Standard 62.1-2007)

A_z = area of the zone

E_z = air distribution effectiveness of the zone

R_p = zone people ventilation rate (5 cfm/person per ASHRAE Standard 62.1-2007)

C_R = measured CO_2 concentration at the room

C_A = measured CO_2 concentration of the outdoor air (OA)

m = metabolic level for occupants of space ($m = 1.2$ for office work)

The above equation has been extracted from *62.1 User's Manual* (ASHRAE 2007b). The control system for the DOAS must be able to calculate the required ventilation rate for each zone based upon the above equation then reset the flow setpoint for the zone to the calculated value. The procedure for a multizone VAV system is somewhat more complicated. The system will require a flow measurement device on the OA inlet for the VAV air handler. The controller must then calculate the actual OA flow to each control zone based upon the percentage of OA in the supply air to the zone. The system must calculate the required OA flow for each zone using the above equation and then must modulate the OA damper to change the supply OA percentage until all zones receive at least the minimum required OA flow.

Selection of CO_2 sensors is critical in both accuracy and response range. CO_2 sensors should be certified by the manufacturer to have an error of 75 ppm or less and be factory calibrated. Inaccurate CO_2 sensors can cause excessive energy use or poor IAQ, so they need to be calibrated as recommended by the manufacturer (see HV18).

Finally, when DCV is used, the system controls should prevent negative building pressure. If the amount of air exhausted remains constant while the intake airflow decreases, the building may be under a negative pressure relative to the outdoors. When air is exhausted directly from the zone (e.g., from restrooms or a janitor's closet), the DCV control strategy must avoid reducing intake airflow below the amount required to replace the air being exhausted.

HV18 *Carbon Dioxide (CO_2) Sensors* **(Climate Zones: all)**

The number and location of CO_2 sensors for DCV can affect the ability of the system to accurately determine the building or zone occupancy. A minimum of one CO_2 sensor per zone is recommended for systems with greater than 500 cfm of OA. Multiple sensors may be necessary if the ventilation system serves spaces with significantly different occupancy expectations. Where multiple sensors are used, the ventilation should be based on the sensor recording the highest concentration of CO_2.

Sensors used in individual spaces should be installed on walls within the space. Multiple spaces with similar occupancies may be represented by an appropriately located sensor in one of the spaces. The number and location of sensors should take into account the sensor manufacturer's recommendations for their particular products as well as the projected usages of the spaces. Sensors should be located such that they provide a representative sampling of the air within the occupied zone of the space. For example, locating a CO_2 sensor directly in the flow path from an air diffuser would provide a misleading reading concerning actual CO_2 levels (and corresponding ventilation rates) experienced by the occupants.

The OA CO_2 concentration can have significant fluctuation in urban areas. $OACO_2$ concentration should be monitored using a CO_2 sensor located near the position of the OA intake. CO_2 sensors should be certified by the manufacturer to have an accuracy to within ± 50 ppm, factory calibrated, and calibrated periodically as recommended by the manufacturer. CO_2 sensors should be calibrated on a regular basis per the manufacturer's recommendations or every six months (per ASHRAE Standard 62.1 [ASHRAE 2007a]).

HV19 *Exhaust Air Systems* **(Climate Zones: all)**

Zone exhaust airflows (for restrooms, janitorial closets, and break rooms) should be determined based on the current version of ASHRAE Standard 62.1 but should not be less than the values required by local code unless approved by the authority having jurisdiction.

Central exhaust systems for restrooms, janitorial closets, and break rooms should be interlocked to operate with the air-conditioning system, except during unoccupied periods. Such a system should have a motorized damper that opens and closes with the operation of the fan. The damper should be located as close as possible to the duct penetration of the building envelope to minimize conductive heat transfer through the duct wall and avoid having to insulate the entire duct. During unoccupied periods, the damper should remain closed and the exhaust fan turned off, even if the air-conditioning system is operating to maintain setback or setup temperatures. Consider designing exhaust ductwork to facilitate recovery of energy (see HV12) from exhaust taken from spaces with air quality classification of 1 or 2 (e.g., restrooms) per Table 6.1 of Standard 62.1 (ASHRAE 2007a).

HV20 *Ductwork Design and Construction* **(Climate Zones: all)**

Low-energy-use ductwork design involves short, direct, and low-pressure-drop runs. The number of fittings should be minimized and should be designed with the least amount of turbulence produced. (In general, the first cost of a duct fitting is approximately the same as 12 ft of straight duct that is the same size as the upstream segment.) Unwanted noise in the ductwork is a direct result of air turbulence. Round duct is preferred over rectangular duct. However, space (height) restrictions may require flat oval ductwork to achieve the low-turbulence qualities of round ductwork. Alternatively, two parallel round ducts may be used to supply the required airflow.

Air should be ducted through low-pressure ductwork with a system pressure classification of less than 2 in. w.c. Rigid ductwork is necessary to maintain low pressure loss and reduce fan energy. Supply air should be ducted to diffusers in each individual space.

In general, the following sizing criteria should be used for duct system components:

- Diffusers and registers, including balancing dampers, should be sized with a static pressure drop no greater than 0.08 in. w.c. Oversized ductwork increases installed cost but reduces energy use due to lower pressure drop.
- Supply ductwork should be sized with a pressure drop no greater than 0.08 in. w.c. per 100 linear feet of duct run. Return ductwork should be sized with a pressure drop no greater than 0.04 in. w.c. and exhaust ductwork with a pressure drop no greater than 0.05 in. w.c.
- Flexible ductwork should be of the insulated type and should be
    - limited to connections between duct branches and diffusers,
    - limited to connections between duct branches and VAV terminal units,
    - limited to 5 ft (fully stretched length) or less,
    - installed without any kinks,
    - installed with a durable elbow support when used as an elbow, and
    - installed with no more than 15% compression from fully stretched length.

Hanging straps, if used, need to use a saddle to avoid crimping the inside cross-sectional area. For ducts 12 in. or smaller in diameter, use a 3 in. saddle; those larger than 12 in. should use a 5 in. saddle.

Long-radius elbows and 45-degree lateral take-offs should be used wherever possible. The angle of a reduction transition should be no more than 45 degrees (if one side is used) or 22.5 degrees (if two sides are used). The angle of expansion transitions should be no more than 15 degrees (laminar air expands approximately 7 degrees).

Air should be returned or exhausted through appropriately placed grilles. Good practice is to direct supply air diffusers toward the exterior envelope and to locate return air grilles near the interior walls, close to the door.

Returning air to a central location (as in a multiple-zone recirculating system) is necessary to reap the benefits of reducing ventilation air due to system population diversity (see HV12). Fully ducted return systems are expensive and must be connected to a single air handler (or the return ducts must be interconnected) to function as a multiple-zone recirculating system. Open plenum return systems are less expensive but must be carefully designed and constructed to prevent infiltration of humid air from the outdoors (Harriman et al. 2001).

The ceiling plenum must also be well sealed to minimize air infiltration. Infiltration can be reduced by using a relief fan to maintain plenum pressure at about 0.05 in. w.c. higher than atmospheric pressure (see HV28), and lowering indoor humidity levels can reduce the risk of condensation (see HV9). In addition, exhaust duct systems should be properly sealed to prevent infiltration.

Caution: Ductwork should not be installed outside the building envelope. Ductwork connected to rooftop units should enter or leave the unit through an insulated roof curb around the perimeter of the unit's footprint. Flexible duct connectors should be used to prevent sound transmission and vibration.

Duct board should be airtight (duct seal level B, from ASHRAE/IES Standard 90.1 [ASHRAE 2010b]) and should be taped and sealed with products that maintain adhesion (such as mastic or foil-based tape). Duct static pressures should be designed and equipment and diffuser selections should be selected not to exceed noise criteria for the space (see HV30 for additional information on noise control).

HV21 **Duct Insulation (Climate Zones: all)**

The following ductwork should be insulated:

- All supply air ductwork
- All return air ductwork located above the ceiling and below the roof
- All OA ductwork
- All exhaust and relief air ductwork between the motor-operated damper and penetration of the building exterior

In addition, all airstream surfaces should be resistant to mold growth and resist erosion, according to the requirements of ASHRAE Standard 62.1 (ASHRAE 2010a).

Exception: In conditioned spaces without a finished ceiling, only the supply air main ducts and major branches should be insulated. Individual branches and run-outs to diffusers in the space being served do not need to be insulated, except where it may be necessary to prevent condensation.

HV22 **Duct Sealing and Leakage Testing (Climate Zones: all)**

The ductwork should be sealed for Seal Class B from ASHRAE/IES Standard 90.1 (ASHRAE 2010b). All duct joints should be inspected to ensure they are properly sealed and insulated, and the ductwork should be leak-tested at the rated pressure. The leakage should not exceed the allowable cubic feet per minute per 100 ft^2 of duct area for the seal and leakage class of the system's air quantity apportioned to each section tested. See HV20 for guidance on ensuring the air system's performance.

HV23 **Fan Motor Efficiencies (Climate Zones: all)**

Motors for fans should meet National Electrical Manufacturers Association (NEMA) premium efficiency motor guidelines (NEMA 2006) when available. Electrically commutated motors may be an appropriate choice for many small units to increase efficiency.

Fan systems should meet or exceed the efficiency levels listed in this chapter and in the recommendation tables in Chapter 4. Depending on the HVAC system type, the efficiency level is expressed in terms of either a maximum power (W) per cubic feet per meter of supply air (for systems where fan power is not included in the packaged HVAC unit efficiency calculation) or a maximum ESP loss (for packaged systems where fan power is included in the EER calculation).

HV24 **Thermal Zoning (Climate Zones: all)**

Office buildings should be divided into thermal zones based on building size, orientation, space layout and function, occupant density, and after-hours use requirements.

Zoning can also be accomplished with multiple HVAC units or a central system that provides independent control for multiple zones. The temperature sensor for each zone should be installed in a location that is representative of the entire zone.

When using a multiple-zone system (such as a VAV system) or a DOAS, avoid using a single air handler (or rooftop unit) to serve zones that have significantly different occupancy patterns. Using multiple air handlers allows air handlers serving unused areas of the building to be shut off, even when another area of the building is still in use. An alternate approach is to use a BAS to define separate operating schedules for these areas of the building, thus shutting off airflow to the unused areas while continuing to provide comfort and ventilation to areas of the building that are still in use.

HV25 **System-Level Control Strategies (Climate Zones: all)**

Control strategies can be designed to help reduce energy. Having a setback temperature for unoccupied periods during the heating season or a setup temperature during the cooling season can help save energy by avoiding the need to operate heating, cooling, and ventilation equipment. Programmable thermostats allow each zone to vary the temperature setpoint based on time of day and day of the week. But they also allow occupants to override these setpoints or ignore the schedule altogether (by using the "hold" feature), which reduces the potential for energy savings. A more sustainable approach is to equip each zone with a zone temperature sensor and then use a system-level controller that coordinates the operation of all components of the system. This system-level controller contains time-of-day schedules that define when different areas of the building are expected to be unoccupied. During these times, the system is shut off and the temperature is allowed to drift away from the occupied setpoint.

A preoccupancy ventilation period can help purge the building of contaminants that build up overnight from the off-gassing of products and packaging materials. Cool temperatures at night can also help precool the building. In humid climates, however, care should be taken to avoid bringing in humid OA during unoccupied periods.

Buildings with multiple-zone recirculating ventilation systems can be designed to account for recirculated OA as well as system population diversity using the equations found in ASHRAE Standard 62.1 (ASHRAE 2010a). In effect, the multiple-zone recirculating ventilation system design approach allows ventilation air to be calculated on the basis of how many people are *in the building* (system population at design) rather than the sum of how many people are *in each space* (sum-of-peak zone population at design). This can reduce the energy required to condition ventilation air in office buildings. Refer to *62.1 User's Manual* (ASHRAE 2007b) for specific guidance.

Optimal start uses a system-level controller to determine the length of time required to bring each zone from the current temperature to the occupied setpoint temperature. Then, the controller waits as long as possible before starting the system so that the temperature in each zone reaches occupied setpoint just in time for occupancy. This strategy reduces the number of hours that the system needs to operate and saves energy by avoiding the need to maintain the indoor temperature at occupied setpoint even though the building is unoccupied.

CHW reset can reduce chiller energy use at part-load conditions. But it should be used only in a constant-flow (not variable-flow) pumping system, and it should be disabled when the outdoor dew point temperature is above 55°F (for example) or if space humidity levels rise to 60% rh or higher.

In a VAV system, SAT reset should be implemented to minimize overall system energy use. This requires considering the trade-off between compressor, reheat, and fan energy as well as the impact on space humidity levels. If SAT reset is used in a humid climate, include one or more zone humidity sensors to disable reset if the relative humidity in the space exceeds 60%.

In order to achieve these ends, control systems should include the following:

- Controls sequences that can easily be commissioned
- A user interface that facilitates understanding and editing of building operating parameters and schedules
- Sensors that are appropriately selected for range of sensitivity and ease of calibration
- Means to effectively convey the current status of systems operation and of exceptional conditions (faults)
- Means to record and convey history of operations, conditions, and efficiencies
- Means to facilitate diagnosis of equipment and systems failures
- Means to document preventive maintenance

See the climate-specific recommendation tables in Chapter 4 for temperature reset strategies in each climate zone.

HV26 *Testing, Adjusting, and Balancing* (**Climate Zones: all**)

After the system has been installed, cleaned, and placed in operation, the system should be tested, adjusted, and balanced in accordance with ASHRAE Standard 111 (ASHRAE 2008b) or SMACNA's testing, adjusting and balancing manual (SMACNA 2002).

Testing, adjusting, and balancing will help to ensure that the correctly sized diffusers, registers, and grilles have been installed, that each space receives the required airflow, and that the fans meet the intended performance. The balancing subcontractor should certify that the instruments used in the measurement have been calibrated within 12 months before use. A written report should be submitted for inclusion in the operation and maintenance (O&M) manuals.

HV27   *Commissioning (Cx)* (Climate Zones: all)

After the system has been installed, cleaned, and placed in operation, the system should be commissioned to ensure that the equipment meets the intended performance and that the controls operate as intended. See Appendix B for more information on commissioning.

HV28   *Filters* (Climate Zones: all)

Particulate air filters are typically included as part of factory-assembled HVAC equipment and should have at least a Minimum Efficiency Reporting Value (MERV) of 6, based on testing procedures described in ASHRAE Standard 52.2 (ASHRAE 2007c).

As explained in *Indoor Air Quality Guide: Best Practices for Design, Construction and Commissioning* (ASHRAE 2009b), EPA maps areas not in compliance (nonattainment) with the National Ambient Air Quality Standards (NAAQS) (EPA 2008a, 2008b). PM2.5 particles are those smaller than 2.5 μm in diameter. In PM2.5 nonattainment areas (virtually all major metropolitan areas), use MERV 11 filters for OA. Use a filter differential pressure gauge to monitor the pressure drop across the filters and send an alarm if the predetermined pressure drop is exceeded. Filters should be replaced when the pressure drop exceeds the filter manufacturer's recommendations for replacement or when visual inspection indicates the need for replacement. The gauge should be checked and the filter should be visually inspected at least once each year.

If high-efficiency filters are to be used, consider using lower-efficiency filters during the construction period. When construction is complete, all filters should be replaced before the building is occupied.

HV29   *Chilled-Water (CHW) System* (Climate Zones: all)

CHW systems efficiently transport cooling energy throughout the building. Often they are combined with thermal storage systems to achieve electrical demand charge savings through mitigation of the peak cooling loads in the building. Thermal storage systems are not covered in this Guide. CHW systems should generally be designed for variable flow through the building.

Small systems (<100 tons) should be designed for variable flow if the chiller unit controls can tolerate expected flow rate changes. CHW systems should use two-way valves with a pressure-controlling bypass set to maintain the minimum evaporator water flow required by the chiller. For chillers that do not tolerate variable CHW flow, three-way valves should be used.

Piping should be sized using the tables in ASHRAE/IES Standard 90.1 (ASHRAE 2010b).

Select cooling coils for a design CHW ΔT of at least 15°F to reduce pump energy. Select cooling coils to minimize air pressure drop. CHW temperature setpoints should be selected based on a life-cycle analysis of pump energy, fan energy, and desired air conditions leaving the coil. Use the recommended temperatures listed in the climate-specific recommendation tables in Chapter 4.

HV30   *Water Heating Systems* (Climate Zones: all)

Condensing boilers can operate at up to 97% efficiency and can operate efficiently at part load. To achieve these high efficiency levels, condensing boilers require that return water temperatures be maintained at 70°F to 120°F, where the boiler efficiency is 97% to 91%. This fits well with hydronic systems that are designed with ΔTs greater than 20°F (the optimal ΔT is 30°F to 40°F). The higher ΔTs allow smaller piping and less pumping energy. Because condensing boilers work efficiently at part load, VFDs can be used on the pumps to further reduce energy use.

Condensing boiler capacity can be modulated to avoid losses caused by cycling at less than full load. This encourages the installation of a modular (or cascade) boiler system, which allows several small units to be installed for the design load but allows the units to match the load for maximum efficiency of the system.

HV31   Relief versus Return Fans (Climate Zones: all)

Relief (rather than return) fans should be used when necessary to maintain building pressurization during economizer operation. Relief fans reduce overall fan energy use in most cases, as long as return dampers are sized correctly. However, if return duct static pressure drop exceeds 0.5 in. w.c., return fans may be needed.

Cautions

HV32   Heating Sources (Climate Zones: all)

Many factors come into play in making a decision whether to use gas or electricity for heating, including availability of service, utility costs, operator familiarity, and the impact of source energy use.

Forced-air electric resistance and gas-fired heaters require a minimum airflow rate to operate safely. These systems, whether stand-alone or incorporated into an air-conditioning or heat pump unit, should include factory-installed controls to shut down the heater when there is inadequate airflow that can result in high temperatures.

Ducts and supply-air diffusers should be selected based on discharge air temperatures and airflow rates.

HV33   Noise Control (Climate Zones: all)

Acoustical requirements may necessitate attenuation of the supply and/or return air, but the impact on fan energy consumption should also be considered and, if possible, compensated for in other duct or fan components. Acoustical concerns may be particularly critical in short, direct runs of ductwork between the fan and supply or return outlet. (See Figure 5-37.)

Avoid installation of the air-conditioning or heat pump units above occupied spaces. Consider locations above less critical spaces such as storage areas, restrooms, corridors, etc. (See Figure 5-38.)

ASHRAE Handbook—HVAC Applications (ASHRAE 2007c) is a potential source for recommended background sound levels in the various spaces that make up office buildings.

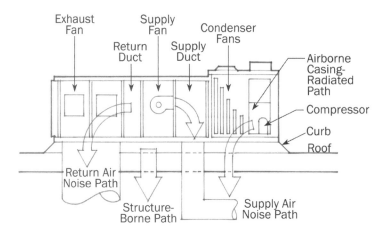

Figure 5-37   (HV33) Typical Noise Paths for Rooftop-Mounted HVAC Units

HV34   *Proper Maintenance* (Climate Zones: all)

Regularly scheduled maintenance is an important part of keeping the HVAC system in optimum working condition. Neglecting preventive maintenance practices can quickly negate any energy savings expected from the system design.

Filters should be replaced when the pressure drop exceeds the filter manufacturer's recommendations for replacement or when visual inspection indicates the need for replacement. ERVs need to be cleaned periodically to maintain performance. Dampers, valves, louvers, and sensors must all be periodically inspected and calibrated to ensure proper operation. This is especially important for OA dampers and CO_2 sensors. Inaccurate CO_2 sensors can cause excessive energy use or poor IAQ, so they need to be calibrated as recommended by the manufacturer.

A BAS can be used to notify O&M staff when preventive maintenance procedures should be performed. This notification can be triggered by calendar dates, run-time hours, the number of times a piece of equipment has started, or sensors installed in the system (such as a pressure switch that indicates when an air filter is too dirty and needs to be replaced).

HV35   *Zone Temperature Control* (Climate Zones: all)

The number of spaces in a zone and the locations of the temperature sensors (thermostats) will affect the control of temperature in the various spaces of a zone. Locating the thermostat in one room of a zone with multiple spaces provides feedback based only on the conditions in that

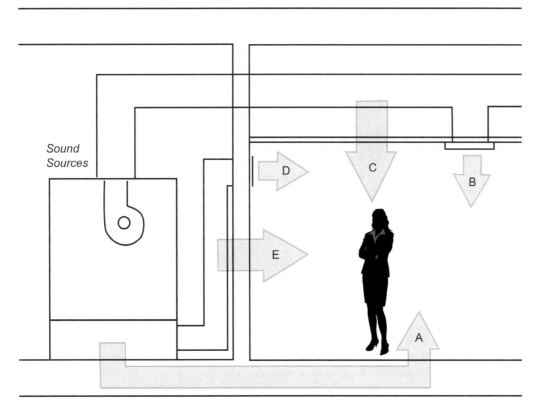

Path A: Structure-borne path through floor
Path B: Airborne path through supply air system
Path C: Duct breakout from supply air duct
Path D: Airborne path through return air system
Path E: Airborne path through mechanical equipment room wall

Figure 5-38   (HV33) Typical Noise Paths for Interior-Mounted HVAC Units

room. Locating a single thermostat in a large open area may provide a better response to the conditions of a zone with multiple spaces. Selecting the room or space that will best represent the thermal characteristics of the space due to both external and internal loads will provide the greatest comfort level for occupants.

To prevent misreading of the space temperature, zone thermostats should not be mounted on exterior walls. Where this is unavoidable, use an insulated sub-base for the thermostat.

In spaces with high ceilings, consider using ceiling fans or high/low air distribution to reduce temperature stratification during the heating season.

Six primary factors must be addressed when defining conditions for thermal comfort:

- Metabolic rate
- Clothing insulation
- Air temperature
- Radiant temperature
- Air speed
- Humidity

Appropriate levels of clothing, the cooling effect of air motion, and radiant cooling or heating systems, for example, can increase occupant comfort while still being energy efficient.

HV36 *Evaporative Condensers on Rooftop Units* (Climate Zones: ❷B ❸B ❹B ❺B)

Evaporative condensers on rooftop DX packaged units can be considered in dry climates to improve energy efficiency. These devices take advantage of the low ambient wet-bulb temperature in order to improve energy efficiency by coupling convective heat rejection with the evaporation of water off of wetted heat rejection condenser coils. In dry climates, up to 40% reduction in energy use can result.

Generally speaking, all of the wetted components and the condenser section should be designed for corrosion resistance to ensure reasonable equipment life. Drawbacks to the system include extra first costs, extra weight that arises from the extra equipment and the water in the sump, additional controls, and the need to provide water treatment regimens.

REFERENCES AND RESOURCES

AHRI. 2007. ANSI/AHRI Standard 340/360-2007, *2007 Standard for Performance Rating of Commercial and Industrial Unitary Air-Conditioning and Heat Pump Equipment.* Arlington, VA: Air-Conditioning, Heating, and Refrigeration Institute.

AHRI. 2008. ANSI/AHRI Standard 210/240, *2008 Standard for Performance Rating of Unitary Air-Conditioning and Air-Source Heat Pump Equipment.* Arlington, VA: Air-Conditioning, Heating, and Refrigeration Institute.

AHRI/ASHRAE. 2005. ANSI/ARI/ASHRAE ISO Standard 13256-1:1998, *Water-source heat pumps—testing and rating for performance—Part 1: Water-to-air and brine-to-air heat pumps.* Atlanta: American Society of Heating, Refrigerating and Air-Conditioning Engineers.

ASHRAE. 2004. ANSI/ASHRAE/IESNA Standard 90.1-2004, *Energy Standard for Buildings Except Low-Rise Residential Buildings.* Atlanta: American Society of Heating, Refrigerating and Air-Conditioning Engineers.

ASHRAE. 2007a. ANSI/ASHRAE Standard 62.1-2007, *Ventilation for Acceptable Indoor Air Quality.* Atlanta: American Society of Heating, Refrigerating and Air-Conditioning Engineers.

ASHRAE. 2007b. *Standard 62.1 User's Manual.* Atlanta: American Society of Heating, Refrigerating and Air-Conditioning Engineers.

ASHRAE. 2007c. ASHRAE Standard 52.2-2007, *Method of Testing General Ventilation Air-Cleaning Devices for Removal Efficiency by Particle Size.* Atlanta: American Society of Heating, Refrigerating and Air-Conditioning Engineers.

ASHRAE. 2007c. *ASHRAE Handbook—HVAC Applications*. Atlanta: American Society of Heating, Refrigerating and Air-Conditioning Engineers.

ASHRAE. 2008a. *ASHRAE Handbook—HVAC Systems and Equipment*. Atlanta: American Society of Heating, Refrigerating and Air-Conditioning Engineers.

ASHRAE. 2008b. ASHRAE Standard 111-1988, *Practices for Measurement, Testing, Adjusting, and Balancing of Building, Heating, Ventilation, Air-Conditioning and Refrigeration Systems*. Atlanta: American Society of Heating, Refrigerating and Air-Conditioning Engineers.

ASHRAE. 2009a. *ASHRAE Handbook—Fundamentals*. Atlanta: American Society of Heating, Refrigerating and Air-Conditioning Engineers.

ASHRAE. 2009b. *Indoor Air Quality Guide: Best Practices for Design, Construction and Commissioning*. Atlanta: American Society of Heating, Refrigerating and Air-Conditioning Engineers.

ASHRAE. 2010a. ANSI/ASHRAE Standard 62.1-2010, *Ventilation for Acceptable Indoor Air Quality*. Atlanta: American Society of Heating, Refrigerating and Air-Conditioning Engineers.

ASHRAE. 2010b. ANSI/ASHRAE/IES Standard 90.1-2010, *Energy Standard for Buildings Except Low-Rise Residential Buildings*. Atlanta: American Society of Heating, Refrigerating and Air-Conditioning Engineers.

ASHRAE. 2010c. *ASHRAE GreenGuide: The Design, Construction, and Operation of Sustainable Buildings*, 3d ed. Atlanta: American Society of Heating, Refrigerating and Air-Conditioning Engineers.

Dieckmann, J., K. Roth, and J. Brodrick. 2003. Dedicated outdoor air systems. *ASHRAE Journal* 45(3):58–59.

EPA. 2008a. National Ambient Air Quality Standards. www.epa.gov/air/criteria.html. Washington, DC: U.S. Environmental Protection Agency.

EPA. 2008b. The Green Book Nonattainment Areas for Criteria Pollutants. www.epa.gov/air/oaqps/greenbk. Washington, DC: U.S. Environmental Protection Agency.

Harriman, L., G. Brundett, and R. Kittler. 2001. *Humidity Control Design Guide for Commercial and Institutional Buildings*. Atlanta: American Society of Heating, Refrigerating and Air-Conditioning Engineers.

Morris, W. 2003. The ABCs of DOAS: Dedicated outdoor air systems. *ASHRAE Journal* 45(5):24–29.

Mumma, S. 2001. Designing dedicated outdoor air systems. *ASHRAE Journal* 43(5):28–31.

Murphy, J. 2006. Smart dedicated outdoor air systems. *ASHRAE Journal* 48(7):30–37.

NEMA. 2006. NEMA Standards Publication MG 1-2006, *Motors and Generators*. Tables 12-12 and 12-13. Rosslyn, VA: National Electrical Manufacturers Association.

Schaffer, M. 2005. *A Practical Guide to Noise and Vibration Control for HVAC Systems* (I-P edition), 2d ed. Atlanta: American Society of Heating, Refrigerating and Air-Conditioning Engineers.

Shank, K., and S. Mumma. 2001. Selecting the supply air conditions for a dedicated outdoor air system working in parallel with distributed sensible cooling terminal equipment. *ASHRAE Transactions* 107(1):562–71.

SMACNA. 2002. *HVAC Systems—Testing, Adjusting and Balancing*, 3d ed. Chantilly, VA: Sheet Metal and Air Conditioning Contractors National Association.

Warden, D. 1996. Dual fan dual duct: Better performance at lower cost. *ASHRAE Journal* 38(1).

QUALITY ASSURANCE

OVERVIEW

Quality assurance (QA) (see also commissioning [Cx]) will help ensure a building functions in accordance with its design intent and thus meets the performance goals established for it. QA should be an integral part of the design and construction processes as well as the continued operation of the facility. General information on QA and Cx is included in Chapter 3, and Appendix C provides examples for the Cx process.

Good Design Practice

QA1   ***Selecting the Design and Construction Team*** **(Climate Zones: all)**

Selection of the design and construction team members is critical to a project's success. Owners need to understand how team dynamics can play a role in the building's resulting performance. Owners should evaluate qualifications of candidates, past performance, costs of services, and availability of the candidates in making their selections. Owners need to be clear in their expectations of how team members should interact. It should be clear that all members should work together to further team goals. The first step is to define members' roles and responsibilities. This includes defining deliverables at each phase during the design and Cx processes.

QA2   ***Selecting the QA Provider*** **(Climate Zones: all)**

QA is a systematic process of verifying the Owner's Project Requirements (OPR), operational needs, and Basis of Design (BoD) and ensuring that the building performs in accordance with these defined needs. The selection of a QA provider should include the same evaluation process the owner would use to select other team members. Qualifications in providing QA services, past performance of projects, cost of services, and availability of the candidate are some of the parameters an owner should investigate and consider when making a selection. Owners may select a member of the design or construction team as the QA provider. While there are exceptions, in general most designers are not comfortable operating and testing assemblies and equipment and most contractors do not have the technical background necessary to evaluate performance. Cx requires in-depth technical knowledge of the building envelope and the mechanical, electrical, and plumbing systems and operational and construction experience. This function is best performed by a third party responsible to the owner because political issues often inhibit a member of the design or construction organizations from fulfilling this responsibility.

QA3   ***Owner's Project Requirements (OPR) and Basis of Design (BoD)***
          (Climate Zones: all)

The OPR details the functional requirements of a project and the expectations of how the facility will be used and operated. This includes strategies and recommendations selected from this Guide (see Chapter 4) that will be incorporated into the project, anticipated hours of operation provided by the owner, and BoD assumptions.

The OPR forms the foundation of the team's tasks by defining project and design goals, measurable performance criteria, owner directives, budgets, schedules, and supporting information in a single, concise document. The QA process depends on a clear, concise, and comprehensive OPR. Development of the OPR document requires input from all key facility users and operators. It is critical to align the complexity of the systems with the capacity and capability of the facility staff.

The next step is for the design team members to document how their design responds to the OPR information. This document is the BoD. It records the standards and regulations, calculations, design criteria, decisions and assumptions, and system descriptions. The narrative must clearly articulate the specific operating parameters required for the systems to form the correct basis for later quality measurements. Essentially, it is the engineering background information that is not provided in the construction documents (CDs) that maps out how the architects and engineers end up with their designs. For example, the BoD would state key criteria such as future expansion and redundancy considerations. It should include important criteria such as what codes, standards, or guidelines are being followed for the various engineered systems, including ventilation and energy. It provides a good place to document owner input needed for engineered systems, such as identifying what electrical loads are to be on emergency power.

QA4   ***Design and Construction Schedule* (Climate Zones: all)**

The inclusion of QA activities in the construction schedule fulfills a critical part of delivering a successful project. Identify the activities and time required for design review and performance verification to minimize time and effort needed to accomplish activities and correct deficiencies.

QA5   ***Design Review* (Climate Zones: all)**

A second pair of eyes provided by the commissioning authority (CxA)/QA provider gives a fresh perspective that allows identification of issues and opportunities to improve the quality of the CDs with verification that the OPR are being met. Issues identified can be more easily corrected early in the project, providing potential savings in construction costs and reducing risk to the team.

QA6   ***Defining QA at Pre-Bid* (Climate Zones: all)**

The building industry has traditionally delivered buildings without using a verification process. Changes in traditional design and construction procedures and practices require education of the construction team that explains how the QA process will affect the various trades bidding the project. It is extremely important that the QA process be reviewed with the bidding contractors to facilitate understanding of and to help minimize fear associated with new practices. Teams who have participated in the Cx process typically appreciate the process because they are able to resolve problems while their manpower and materials are still on the project, significantly reducing delays, callbacks, and associated costs while enhancing their delivery capacity.

These requirements can be reviewed by the architect and engineer of record at the pre-bid meeting, as defined in the specifications.

QA7   ***Verifying Building Envelope Construction* (Climate Zones: all)**

The building envelope is a key element of an energy-efficient design. Compromises in assembly performance are common and are caused by a variety of factors that can easily be avoided. Improper placement of insulation, improper sealing or lack of sealing at air barriers, wrong or poorly performing glazing and fenestration systems, incorrect placement of shading devices, misplacement of daylighting shelves, and misinterpretation of assembly details can significantly compromise the energy performance of the building (see the "Cautions" sections throughout this chapter). The perceived value of the Cx process is that it is an extension of the quality control processes of the designer and contractor as the team works together to produce quality energy-efficient projects.

QA8   ***Verifying Lighting Construction* (Climate Zones: all)**

Lighting plays a significant role in the energy consumption of the building. Lighting for the all of the space types should be reviewed against anticipated schedules of use throughout the day.

QA9 *Verifying Electrical and HVAC Systems Construction* (Climate Zones: all)

Performance of electrical and HVAC systems is a key element of this Guide. How systems are designed as well as installed affect how efficiently they will perform. Collaboration between the entire design team is needed to optimize the energy efficiency of the facility. Natural daylight and artificial lighting will impact the heating and cooling loads, resulting in impacts on both HVAC equipment capacity and hourly operation mode. The design reviews should pay close attention to electrical and HVAC systems. Proper installation is just as important as proper design. Making sure the installing contractor's foremen understand the owner's goals, the QA process, and the installation details is a key factor to system performance success. A significant part of this process is a careful and thorough review of product submittals to ensure compliance with the design. It is in everyone's best interest to install the components correctly and completely the first time. Trying to inspect quality once a project is already under construction is time consuming, costly, and usually doesn't result in quality. It's much better to ensure all team members are aligned with the QA process and goals. Certainly, observations and inspections during construction are necessary. Timing is critical to ensure that problems are identified at the beginning of each system installation. This minimizes the number of changes (saving time and cost) and leaves time for corrections.

QA10 *Performance Testing* (Climate Zones: all)

Performance testing of systems is essential to ensure that all the commissioned systems are functioning properly in all modes of operation. This is a prerequisite for the owner to realize the energy savings that can be expected from the strategies and recommendations contained in this Guide. Unlike most appliances these days, none of the mechanical/electrical systems in a new facility are "plug and play." If the team has executed the Commissioning Plan and is aligned with the QA goals, the performance testing will occur quickly and only minor issues will need to be resolved. Owners with O&M personnel can use the functional testing process as a training tool to educate their staff on how the systems operate as well as for system orientation prior to training.

QA11 *Substantial Completion* (Climate Zones: all)

Substantial completion generally means the completion and acceptance of the life safety systems and that the facility is ready to be occupied. All of the systems should be operating as intended. Expected performance can only be accomplished when all systems operate interactively to provide the desired results. As contractors finish their work, they will identify and resolve many performance problems. The CxA/QA provider verifies that the contractor maintained a quality control process by directing and witnessing testing and then helps to resolve remaining issues.

QA12 *Final Acceptance* (Climate Zones: all)

Final acceptance generally occurs after the Cx/QA issues in the issues log have been resolved, except for minor issues the owner is comfortable with resolving during the warranty period.

QA13 *Establish a Building Operation and Maintenance (O&M) Program* **(Climate Zones: all)**

Continued performance and control of operational and maintenance costs require a maintenance program. The O&M manuals provide information that the O&M staff uses to develop this program. Detailed O&M system manual and training requirements are defined in the OPR and executed by the project team to ensure the O&M staff has the tools and skills necessary. The level of expertise typically associated with O&M staff for buildings covered by this Guide is generally much lower than that of a degreed or licensed engineer, and they typically need assistance with development of a preventive maintenance program. The CxA/QA provider can help bridge the knowledge gaps of the O&M staff and assist the owner with developing a program that helps ensure continued performance. The benefits associated with energy-efficient buildings are realized when systems perform as intended through proper design, construction, operation, and maintenance.

QA14   *Monitor Post-Occupancy Performance* **(Climate Zones: all)**

Establishing measurement and verification procedures for actual building performance after a building has been commissioned can identify when corrective action and/or repair is required to maintain energy performance. Utility consumption and factors affecting utility consumption should be monitored and recorded to establish building performance during the first year of operation.

Variations in utility usage can be justified based on changes in conditions typically affecting energy use, such as weather, occupancy, operational schedule, maintenance procedures, and equipment operations required by these conditions. While most buildings covered in this Guide will not use a formal measurement and verification process, tracking the specific parameters listed above does allow the owner to quickly review utility bills and changes in conditions. Poor performance is generally obvious to the reviewer when comparing the various parameters. CxA/QA providers can typically help owners understand when operational tolerances are exceeded and can provide assistance in defining what actions may be required to return the building to peak performance.

Another purpose of the post-occupancy evaluation is to determine the actual energy performance of the current generation of low-energy buildings to verify design goals and document real-world energy savings. Additionally, post-occupancy evaluations provide lessons learned in the design, technologies, operation, and analysis techniques to ensure these and future buildings operate at a high level of performance over time. For details and some case studies and lessons learned, refer to the published report by National Renewable Energy Laboratory (NREL) (Torcellini et al. 2006).

REFERENCES AND RESOURCES

ASHRAE. 2005. ASHRAE Guideline 0-2005, *The Commissioning Process*. Atlanta: American Society of Heating, Refrigerating and Air-Conditioning Engineers.

ASHRAE. 2007. ASHRAE Guideline 1.1-2007, *HVAC&R Technical Requirements for The Commissioning Process*. Atlanta: American Society of Heating, Refrigerating and Air-Conditioning Engineers.

Torcellini, P., S. Pless, M. Deru, B. Griffith, N. Long, and R. Judkoff. 2006. Lessons learned from case studies of six high-performance buildings, Technical Report NRETL/TP-55-037542. www.nrel.gov/docs/fy06osti/37542.pdf. Golden, CO: National Renewable Energy Laboratory.

ADDITIONAL BONUS SAVINGS

DAYLIGHTING—TOPLIGHTING

DL21   *Toplighting* (Climate Zones: all)

Toplighting draws from zenithal skylight, which makes toplighting the most effective source of daylight. Toplighting therefore requires smaller apertures than sidelighting to achieve the same level of light. In office buildings, toplighting is recommended for use in occupied spaces that have no access to sidelight. Toplighting is best used in circulation areas and contiguous spaces that are used for reception areas or lobbies. Toplighting in circulation areas needs careful coordination with overhead ductwork and lighting but does not limit future flexibility as required in program spaces.

Toplighting is a highly effective strategy that not only provides excellent daylight and way-finding support but also can save energy for electrical lighting and cooling. The limitation of toplighting is that it can be used in single-story designs only or on the top floors of multistory designs. Two types of toplighting can be distinguished, as noted in DL22 and DL24.

DL22   *Rooftop Monitors* (Climate Zones: all)

Rooftop monitors are typically the toplighting strategy best suited for office building applications. The monitor's vertical glazing delivers excellent quality daylight and delivers it specifically to the monitor's orientation (which is important for good controlling of the daylight). Roof monitors should not face east or west. South orientation is possible if appropriately sized overhangs are included, but undesired solar heat gain is blocked most effectively when the monitors face north. (See Figure 5-39.)

Fenestration to Floor Area Ratio (FFR) of Rooftop Monitors. A 10% fenestration to floor area ratio (FFR) of vertical glazing with a VT in accordance with the values for vertical fenestration in the climate-specific recommendation tables in Chapter 4 is sufficient to achieve good quality daylight levels and to switch off electrical lighting during daytime in all climates

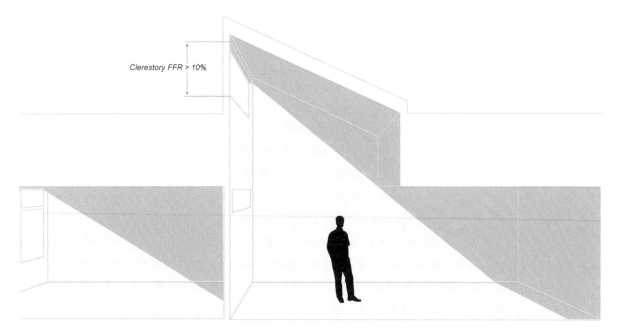

Clerestory FFR > 10%

Figure 5-39   (DL22) Rooftop Monitor

and under partially cloudy sky conditions. When the monitor faces south, the glazing area is typically 25% less than when it faces north to provide the same amount of daylighting.

Rooftop monitors add volume. In spaces using all-air system environments with a cubic foot per minute rate based on square footage, the added volume should be taken into consideration, as the volume increase can lead to a higher cubic foot per minute rate and incur higher energy consumption.

DL23   ***Rooftop Monitor Design* (Climate Zones: all)**

To help reduce conductive gains and losses, the walls and ceilings of the roof monitor should be insulated and should incorporate appropriate insulation barriers as recommended in EN2 and EN6. Make sure that the colors used within the monitor are light and comply with the minimum reflectances in Table 5-3 of this chapter. White works best. Darker colors will result in a considerable loss of efficiency.

Also consider acoustic issues. If acoustical ceiling material is used, make sure that the reflectance and the acoustical properties are high. Often, in presenting the reflectance of an acoustical tile, manufacturers will specify the reflectance of the paint. Remember to account for reduced reflectance caused by the fissures in the tile. (See Figure 5-22.)

DL24   ***Skylights* (Climate Zones: ⑤ ⑥ ❼ ❽)**

Skylights are a powerful source of daylight; however, their difficulties with handling solar heat gain, direct beam radiation, and glare pose challenges when designing for demanding work environments. The maximum skylight fenestration area, the amount of skylight opening to gross roof area, should not exceed 5%. Applications in office buildings should be considered with great care and only if north-facing rooftop monitors are not deemed feasible.

In general, the shape and interior wall reflectances of skylight wells limit the useful amount of daylight transmitted into the room. Interior reflectances should be above 70%. In regard to proportions, a good rule of thumb is that the well depth should be less than twice the well width.

Work spaces need to be shielded from direct sun. Diffusing skylights can cause glare. Use light-reflecting baffles and/or diffusing glazing to control direct sun. (See Figure 5-40.)

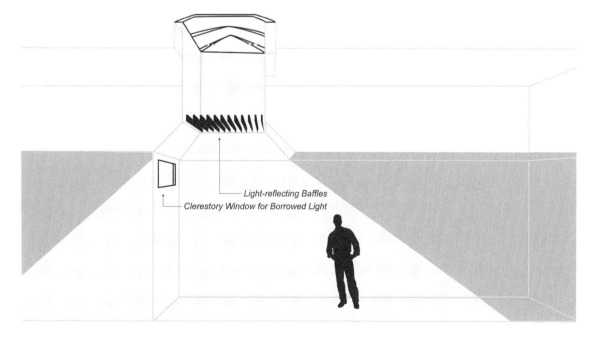

Light-reflecting Baffles
Clerestory Window for Borrowed Light

Figure 5-40   (DL24) Roof Skylight and Space Section

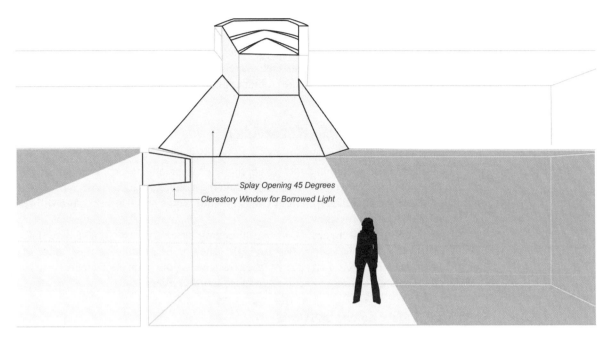

Splay Opening 45 Degrees
Clerestory Window for Borrowed Light

Figure 5-41   (DL26) Roof Skylight Section

Reduce thermal gain during the cooling season by using skylights with a low overall thermal transmittance (U-factor). Insulate the skylight curb above the roof line with rigid continuous installation (c.i.). Shade skylights with exterior/interior sun control devices such as screens, baffles, or fins.

DL25   Toplighting—Thermal Transmittance (**Climate Zones: ❶ ❷ ❸**)

In hot climates, use north-facing monitors for toplighting whenever possible to eliminate excessive solar heat gain and glare. Typically, north-facing monitors have one-sixth the heat gain of skylights.

Reduce thermal gain during the cooling season by using skylights with a low overall thermal transmittance (U-factor). Insulate the skylight curb above the roof line with rigid c.i. Shade skylights with exterior/interior sun control devices such as screens, baffles, or fins.

DL26   Toplighting—Thermal Transmittance (**Climate Zones: ④ ❺ ❻ ❼ ❽**)

In moderate and cooler climates, use either north- or south-facing rooftop monitors for toplighting but not east- or west-facing monitors. East-west glazing adds excessive summer heat gain and makes it difficult to control direct solar gain. Monitors with operable glazing may also help provide natural ventilation in temperate seasons when air conditioning is not in use.

Reduce summer heat gain as well as winter heat loss by using skylights with a low overall thermal transmittance (U-factor) and by controlling skylight fenestration area. Use a skylight frame that has a thermal break to prevent excessive heat loss/gain and winter moisture condensation on the frame. Insulate the skylight curb above the roof line with rigid c.i.

Shade south-facing rooftop monitors and skylights with exterior/interior sun control devices such as screens, baffles, or fins. As shown in Figure 5-41, splay the skylight opening at 45° to maximize daylight distribution and minimize glare.

DL27   Toplighting—Ceiling Height Differentials (**Climate Zones: all**)

Differences in floor-to-floor height offer useful opportunities for daylighting. Differentials in ceiling height as a result of programmatic requirements provide cost-effective opportunities to implement daylight through toplighting. (See Figure 5-42.)

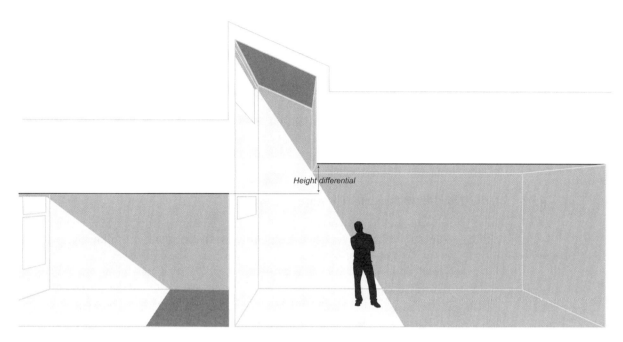

Figure 5-42 (DL27) Toplighting Height Differential—South-Facing

NATURAL VENTILATION

NV1   *Natural Ventilation and Naturally Conditioned Spaces*
(Climate Zones: ❷B ❸B ❸C   ❺B and as appropriate elsewhere in a mixed-mode approach)

Natural ventilation involves the use of operable elements in the façade of a building to bring in OA. ASHRAE Standard 62.1 requires either engineering analysis to confirm adequate ventilation effectiveness or compliance with prescriptive requirements that govern the size and spacing of the openings, as well as permanent accessibility to the controls by the occupants (ASHRAE 2010a).

Occupant-controlled naturally conditioned spaces, as defined by ASHRAE Standard 55, are "those spaces where the thermal conditions of the space are regulated primarily by the opening and closing of windows by the occupants" (ASHRAE 2010b, p. 3). Standard 55 allows an adaptive comfort standard to be used under a limited set of conditions.

When considering either natural ventilation or naturally conditioned spaces, one must first consider the climate and the number of hours when an occupant might want to open the windows and evaluate whether a natural scheme is possible for the range of outdoor temperatures and humidities. If the climate performance supports natural ventilation/conditioning, then the designer should also review the OA quality (as OA pollutants are generally not filtered and OA quality will heavily influence IAQ), the noise impacts from adjacent streets, and security concerns.

A mixed-mode approach usually falls into one of three categories (CBE 2005):

- *Zoned use*—one area of the building is naturally conditioned/ventilated while another is fully air conditioned
- *Change-over use*—the area is naturally ventilated and conditioned for a portion of the year but is fully heated and cooled during extreme seasons
- *Concurrent use*—the area is naturally ventilated but artificially cooled (usually through a passive radiant system)

It is important to evaluate the frequency of natural ventilation use because in mixed-mode systems the owner is often purchasing two systems (an air-conditioning system and the operable windows in the façade).

RENEWABLE ENERGY

RE1 *Photovoltaic (PV) Systems* **(Climate Zones: all)**

Photovoltaic (PV) systems have become an increasingly popular option for on-site electric energy production. These systems require very little maintenance and generally have long lifetimes.

Options for installing PV systems include on rooftops (including collectors integrated with the roofing membrane), ground mounted, or as the top of a covered parking system. The systems may be fixed-mounted or tracking. Each installation method offers different combinations of advantages and disadvantages.

RE2 *Solar Hot Water Systems* **(Climate Zones: all)**

Simple solar systems are most efficient when they generate heat at low temperatures. General suggestions for solar domestic hot water heating systems include the following:

- It is typically not economical to design solar systems to satisfy the full annual SWH load.
- Systems are typically most economical if they furnish 50%–80% of the annual load.
- Properly sized systems will meet the full load on the *best* solar day of the year.
- Approximately 1–2 gal of storage should be provided per square foot of collector.
- In general, 1 ft^2 of collector heats about 1 gal/day of service water at 44° latitude.
- Glazed flat-plate systems often cost in the range of $100 to $150/ft^2 of collector.
- Collectors do not have to face due south; they receive 94% of the maximum annual solar energy if they are 45° east or west of due south.
- The optimal collector tilt for service water applications is approximately equal to the latitude where the building is located; however, variations of ±20° only reduce the total energy collected about 5%. This is one reason that many collector installations are flat to a pitched roof instead of being supported on stands.
- The optimal collector tilt for building heating (not domestic water heating) systems is approximately the latitude of the building plus 15°.

Collectors can still function on cloudy days to varying degrees depending on the design, but they perform better in direct sunlight; collectors should not be placed in areas that are frequently shaded.

Solar systems in most climates require freeze protection. The two common types of freeze protection are drainback systems and systems that contain antifreeze.

Drainback solar hot water systems are often selected in small applications where the piping can be sloped toward a collection tank. By draining the collection loop, freeze protection is accomplished when the pump shuts down, either intentionally or unintentionally. This avoids the heat transfer penalties of antifreeze solutions.

Closed-loop, freeze-resistant solar systems should be used when piping layouts make drainback systems impractical.

In both systems, a pump circulates water or antifreeze solution through the collection loop when there is adequate solar radiation and a need for service water heat.

Solar collectors for service water applications are usually flat-plate or evacuated-tube type. Flat-plate units are typically less expensive. Evacuated-tube designs can produce higher temperatures because they have less standby loss but also can pack with snow and if fluid flow stops are more likely to reach temperatures that can degrade antifreeze solutions.

Annual savings can be estimated using performance data from the Solar Rating and Certification Corporation Web site (SRCC 2011). A free downloadable program called RETScreen from Natural Resources Canada (NRCan 2010) can assist with economic feasibility analysis, and many utility rebate programs use it in calculating rebates or determining eligibility. The first cost of the system must be estimated.

Using Solar Energy

PV Panels
Source: Cody Andresen, Arup

Evacuated-Tube Solar Heaters
Source: Gerard Healey

Where solar energy is prevalent, it can be used to create electricity from PV panels or hot water from evacuated-tube solar heaters.

Photovoltaics are now commonplace in most markets as a renewable energy source, with many municipalities and utilities allowing a net-metering approach that allows the building owner to push electricity into the grid during daytime hours and remove it at other times as needed. There is little maintenance other than cleaning, which is required for PV panels.

Evacuated-tube solar heaters are a more complex option for the use of solar energy, as they require that a base load of heating is required throughout the year. Most designers wishing to incorporate solar heaters should consult a specialist regarding the specifics of the installation.

RE3   Wind Turbine Power (**Climate Zones: all**)

Wind energy is one of the lowest-priced renewable energy technologies available today, costing between 5 to 11 cents per kilowatt-hour, depending upon the wind resource and project financing of the particular project. Small- to medium-sized wind turbines are typically considered for small to medium office buildings. These turbines range from 4 to 200 kW and are typically mounted on towers from 50 to 100 ft and connected to the utility grid through the building's electrical distribution system.

One of the first steps to developing a wind energy project is to assess the area's wind resources and estimate the available energy. From wind resource maps, you can determine if your area of interest should be further explored. Note that the wind resource at a micro level can vary significantly; therefore, you should get a professional evaluation of your specific area of interest.

The map in Figure 5-43 shows the annual average wind power estimates at 50 m above ground. It combines high- and low-resolution data sets that have been screened to eliminate land-based areas unlikely to be developed due to land use or environmental issues. In many states, the wind resource has been visually enhanced to better show the distribution on ridge crests and other features. Estimates of the wind resource are expressed in wind power classes ranging from Class 1 (lowest) to Class 7 (highest), with each class representing a range of mean wind power density or equivalent mean speed at specified heights above the ground. This map does not show Classes 1 and 2, as Class 2 areas are marginal and Class 1 areas are unsuitable for wind energy development. In general, at 50 m, Class 4 or higher wind power can be useful for generating wind power. More detailed state wind maps are available at the U.S. Department of Energy's Office of Energy Efficiency and Renewable Energy Web site (EERE 2011a).

Although wind turbines themselves do not take up a significant amount of space, they need to be installed an adequate distance from the nearest building for several reasons, including turbulence reduction (which affects efficiency), noise control, and safety. It is essential that coordination occurs between the owner, design team, and site planner to establish the optimal wind turbine location relative to the other facilities on the site.

The three largest complaints about wind turbines are the noise, the killing of birds, and the aesthetic appearance. Most of these problems have been resolved or greatly reduced through technological developments or by properly siting wind turbines. Most small wind turbines today have an excellent safety record. An important factor is to consider how the wind turbine controls itself and shuts itself down. Can operators shut it off and stop the turbine when they want or need to do so? This is extremely important, and unfortunately there are very few small turbines that have reliable means to stop the rotor on command. The few that do may require you to do so from the base of the tower—not exactly where you want to be if the turbine is out of control in a wind storm. Look for a system that offers one or more means to shut down and preferably stop the rotor remotely.

Using energy modeling, the electric energy consumption of the building can be modeled. Using this data in conjunction with the financial details of the project, including the rebates, the owner and designer must then chose the correct size turbine that meets their needs. Note that the closer the match of the turbine energy output to the demand, the more cost-effective the system will be. Make sure that all costs are listed to give a total cost of ownership for the wind turbine. This includes the wind turbine, the tower, the electrical interconnection, controls, installation, maintenance, concrete footings, guy wires, and cabling.

In addition to evaluating the initial cost of the turbine, it is extremely important to consider the federal and state policies and incentive programs that are available. The Database for State Incentives for Renewables and Efficiency (DSIRE) provides a list of available incentives, grants, and rebates (EERE 2011b). Also critical to the financial success to a wind turbine project is a favorable net metering agreement with the utility.

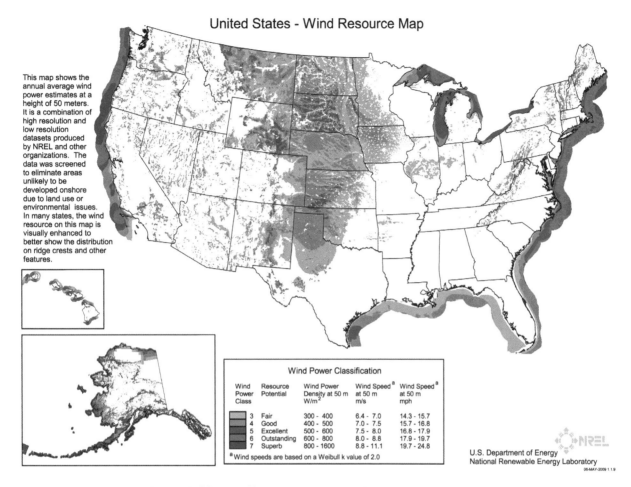

Figure 5-43   (RE3) Average Annual Wind Power Estimates
Source: NREL (2009)

REFERENCES

ASHRAE. 2010a. ANSI/ASHRAE Standard 62.1-2010, *Ventilation for Acceptable Indoor Air Quality*. Atlanta: American Society of Heating, Refrigerating and Air-Conditioning.

ASHRAE. 2010b. ANSI/ASHRAE Standard 55-2010, *Thermal Environmental Conditions for Human Occupancy*. Atlanta: American Society of Heating, Refrigerating and Air-Conditioning Engineers.

EERE. 2011a. Wind Powering America, Wind and Water Program. www.windpoweringamerica.gov/wind_maps.asp. Washington, DC: U.S. Department of Energy, Office of Energy Efficiency and Renewable Energy.

EERE. 2011b. Database for State Incentives for Renewables and Efficiency (DSIRE). www.dsireusa.org. Washington, DC: U.S. Department of Energy, Office of Energy Efficiency and Renewable Energy.

NRCan. 2010. RETScreen from Natural Resources Canada (www.retscreen.net)

NREL. 2009. Wind Resources (50 m) of the United States. Image "map_wind_national" available at www.nrel.gov/gis/images. Golden, CO: National Renewable Energy Laboratory.

SRCC. 2011. Ratings. www.solar-rating.org/ratings/ratings.htm. Cocoa, FL: Solar Rating and Certification Corporation.

CBE. 2005. About Mixed Mode. Center for the Built Environment Web site, www.cbe.berkeley.edu/mixedmode/aboutmm.html. Berkeley: University of California–Berkeley.

Appendix A— Envelope Thermal Performance Factors

Each recommendation table in Chapter 4 presents a prescriptive construction option for each opaque envelope measure. Table A-1 presents U-factors for above-grade components, C-factors for below-grade walls, and F-factors for slab-on-grade floors that correspond to each prescriptive construction option. Alternative constructions would be equivalent methods for meeting the recommendations of this Guide provided they are less than or equal to the thermal performance factors listed in Table A-1.

Table A-1 Opaque Construction Options

Roof Assemblies		Walls, Above Grade		Floors	
R	**U**	**R**	**U**	**R**	**C**
Insulation Above Deck		**Mass Walls**		**Mass**	
20	0.048	5.7	0.151	4.2 c.i.	0.137
25	0.039	7.6	0.123	10.4 c.i.	0.074
30	0.032	11.4	0.090	12.5 c.i.	0.064
35	0.028	13.3	0.080	14.6 c.i.	0.056
Attic and Other		19.0	0.066	16.7 c.i.	0.051
38	0.027	**Steel Framed**		20.9 c.i.	0.042
49	0.021	13 + 7.5 c.i.	0.064	23.0 c.i.	0.038
60	0.017	13 + 15.6 c.i.	0.042	**Steel Framed**	
Metal Building		13 + 18.8 c.i.	0.037	19	0.052
19 + 10 FC	0.057	**Wood Framed and Other**		13 + 7.5 c.i.	0.045
19 + 11 Ls	0.035	13	0.089	30	0.038
25 + 11 Ls	0.031	13 + 3.8 c.i.	0.064	38	0.032
30 + 11 Ls	0.029	13 + 7.5 c.i.	0.051	49	0.027
25+11+11 Ls	0.026	13 + 10.0 c.i.	0.045	60	0.024
		13 + 12.5 c.i.	0.040	**Wood Framed and Other**	
		13 + 15.0 c.i.	0.037	19	0.051
Slabs		13 + 18.9 c.i.	0.032	30	0.033
R - in.	**F**	**Metal Building**		38	0.027
Unheated		0 + 9.8 c.i.	0.094	49	0.022
10 - 24	0.54	0 + 13.0 c.i.	0.072	60	0.018
20 - 24	0.51	0 + 15.8 c.i.	0.060		
Heated		0 + 19.0 c.i.	0.050		
7.5 - 12	1.02	0 + 22.1 c.i.	0.044		
10 - 24	0.90	0 + 25.0 c.i.	0.039		
15 - 24	0.86				
20 - 24	0.843				
20 - 48	0.688	**Walls, Below Grade**			
25 - 48	0.671	**R**	**C**		
20 full slab	0.373	7.5 c.i.	0.119		
		10.0 c.i.	0.092		
		15.0 c.i.	0.067		

C = thermal conductance, Btu/h·ft^2·°F
c.i. = continuous insulation
F = slab edge heat loss coefficient per foot of perimeter, Btu/h·ft·°F
FC = filled cavity
Ls = liner system
R = thermal resistance, h·ft^2·°F/Btu
R - in. = R-value followed by the depth of insulation in inches
U = thermal transmittance, Btu/h·ft^2·°F

Appendix B— International Climatic Zone Definitions

Table B-1 shows the climate zone definitions that are applicable to any location. The information is from ASHRAE/IESNA Standard 90.1-2007, Normative Appendix B, Table B-4 (ASHRAE 2007). Weather data is needed in order to use the climate zone definitions for a particular city. Weather data by city is available for a large number of international cities in 2009 *ASHRAE Handbook—Fundamentals* (ASHRAE 2009).

Table B-1   International Climatic Zone Definitions

Climate Zone Number	Name	Thermal Criteria*
1A and 1B	Very Hot–Humid (1A) Dry (1B)	9000 < CDD50°F
2A and 2B	Hot–Humid (2A) Dry (2B)	6300 < CDD50°F ≤ 9000
3A and 3B	Warm–Humid (3A) Dry (3B)	4500 < CDD50°F ≤ 6300
3C	Warm–Marine (3C)	CDD50°F ≤ 4500 AND HDD65°F ≤ 3600
4A and 4B	Mixed–Humid (4A) Dry (4B)	CDD50°F ≤ 4500 AND 3600 < HDD65°F ≤ 5400
4C	Mixed–Marine (4C)	3600 < HDD65°F ≤ 5400
5A, 5B, and 5C	Cool–Humid (5A) Dry (5B) Marine (5C)	5400 < HDD65°F ≤ 7200
6A and 6B	Cold–Humid (6A) Dry (6B)	7200 < HDD65°F ≤ 9000
7	Very Cold	9000 < HDD65°F ≤ 12600
8	Subarctic	12600 < HDD65°F

*CDD = cooling degree-day, HDD = heating degree-day.

DEFINITIONS

Marine (C) Definition—Locations meeting all four of the following criteria:

- Mean temperature of coldest month between 27°F and 65°F
- Warmest month mean < 72°F
- At least four months with mean temperatures over 50°F
- Dry season in summer. The month with the heaviest precipitation in the cold season has at least three times as much precipitation as the month with the least precipitation in the rest of the year. The cold season is October through March in the Northern Hemisphere and April through September in the Southern Hemisphere.

Dry (B) Definition—Locations meeting the following criterion:

- Not marine and $P < 0.44 \times (T - 19.5)$
  where
  P = annual precipitation, in.
  T = annual mean temperature, °F

Moist (A) Definition—Locations that are not marine and not dry.

REFERENCES

ASHRAE. 2007. ANSI/ASHRAE/IESNA Standard 90.1-2007, *Energy Standard for Buildings Except Low-Rise Residential Buildings*. Atlanta: American Society of Heating, Refrigerating and Air-Conditioning Engineers.

ASHRAE. 2009. *ASHRAE Handbook—Fundamentals*. Atlanta: American Society of Heating, Refrigerating and Air-Conditioning Engineers. [Available in print form and on CD-ROM.]

Appendix C— Commissioning Information and Examples

Following are examples of what a commissioning scope of services and a responsibility matrix (Table C-1) might include. Project teams should adjust these to meet the needs of the owner and project scope, budget, and expectations.

COMMISSIONING SCOPE OF SERVICES

INTRODUCTION

Commissioning (Cx) is a quality assurance (QA) process with four main elements. First, the architectural and engineering team must clearly understand the building owner's goals and requirements for the project. Next the architectural and engineering team must design systems that support or respond to those requirements. The construction team must understand how the components of the system must come together to ensure that the system is installed correctly and performs as intended. Last, the operators of the system must also understand how the system is intended to function and have access to information that allows them to maintain it as such. This process requires more coordination, collaboration, and documentation between project team members than traditionally has been provided.

The intent of this appendix is to help provide an understanding of the tasks, deliverables, and costs involved. An independent commissioning authority (CxA), one that is contracted directly with the building owner, will be the building owner's representative to facilitate the Cx process and all of its associated tasks. The CxA will lead the team to ensure everyone understands the various tasks, the roles they play, and the desired outcome or benefit for following the Cx process. The systems required to be commissioned are those that impact the use of energy. Project team members responsible for the design or installation of those systems will have the majority of the Cx work. The majority of the field work will be the responsibility of the mechanical, electrical, and control contractors.

Cx of a new building will ultimately enhance the operation of the building. Reduced utility bills, lower maintenance costs, and a more comfortable and healthier indoor environment will result. Cx focuses on creating buildings that are as close to the owners' and users' objectives (as delineated in the Owner's Project Requirements) as possible. Early detection and resolution of potential issues are the keys to achieving a high-quality building without increasing the total effort and cost to the team members. Resolving design issues early will significantly reduce efforts during construction. Finding mistakes after installation or during start-up are costly to everyone. Checklists will assist the contractors during installation, and installation issues will be detected early. Early detection will reduce the amount of rework required compared to late detection at final inspection. This will also benefit the owner and occupants since the building will work as intended from day one of operation.

SYSTEMS

The systems under this scope of services include the following.

- The entire heating, ventilating, and air-conditioning (HVAC) system (boilers, chillers, pumps, piping, and air distribution systems).
- The building automation system (BAS) for the HVAC system.
- The domestic hot-water system.
- The electrical systems (lighting and receptacle systems, electrical panels, transformers, motor control centers, electrical motors, and other electrical items excluding emergency power systems).
- The building envelope as it relates to energy efficiency (insulation, wall framing—thermal bridging, air leakage, glazing solar and thermal characteristics, and fenestration framing—thermal bridging).

These listed systems will be commissioned by the tasks described in the "Commissioning Tasks" section of this chapter.

DELIVERABLES

The following deliverables are part of the Cx scope of services.

- Commissioning Plan
- Owner's Project Requirements (OPR)
- CxA's design review
- Construction installation checklists
- CxA's site visit reports
- System functional performance tests
- Systems manual
- Owner training
- Cx report
- Systems warranty review
- Final Cx report

SCHEDULE

Cx begins in the early stage of design and continues through building operation. The following sections detail the specific step-by-step activities that owners, designers, and construction team members need to follow in each phase of the project's delivery.

Planning Phase

- Document OPR (project intent)
- Develop Commissioning Plan
- Specify architect/engineer Cx requirements
- Assist with the architect/engineer selection process

Design Phase

- Verify that the design meets the OPR
- Write Cx specifications

Construction Phase

- Verify that the submittals meet OPR
- Verify that the installation meets OPR
- Verify that the components function as required
- Facilitate training of building operators

Acceptance Phase

- Verify that the systems work as required and meets OPR
- Verify that the OPR are met throughout the building
- Develop systems manual
- Review contractors' operation and maintenance and systems manuals

Operational Phase

- Warranty review
- Verify that the operation of the building is optimal

COMMISSIONING TASKS

Commissioning Plan

The CxA will write the Commissioning Plan and detail the Cx tasks and schedule for executing the Commissioning Plan tasks. In addition, the communication channels will be listed and samples of all forms, procedures, and checklists used for the project will be provided. The Cx responsibilities of each of the project team members will be listed. The Commissioning Plan will be updated as the project progresses and as forms, procedures, checklists, schedules, agendas, reports, etc., are finalized or revised. These updates will be distributed at major milestones to all project team members.

Owner's Project Requirements (OPR)

The OPR document describes the main expectations the owner wants the project to meet. As the owner usually wants to meet most of the expectations of all stakeholders, input from a representative of each stakeholder is beneficiary.

For the referenced project the CxA will facilitate and write the OPR with input from the owner.

Commissioning Specifications

The Cx specifications will clearly state what will be expected from the contractors. This will include activities the contractor needs to participate in and documentation procedures required throughout the construction period. Sample forms and procedures will be provided to show the contractors visuals of what they will need to complete in the construction and acceptance phases. The Cx specifications will also include the training requirements as well as the documentation needed to develop the systems manual.

The CxA will provide the requirements for Cx in the construction phase to be integrated into the specifications.

Basis of Design (BoD)

The Basis of Design (BoD) includes all engineering and architectural calculations and assumptions on how to design the systems such that the OPR are met. This document will be written by the architect and engineers and will be reviewed by the CxA for completeness and quality. Comments will be provided if any pertinent information is missing or if more details are needed. The BoD will need to be updated if any changes occur during the project. This is needed to inform all project team members about revised assumptions and new directions the project is heading in.

Design Review

During the design review the CxA will focus on verifying that the OPR will be met. In addition, the design documents will be reviewed for constructability, operability, and maintainability. The review will take place at 70% completion and be back-checked for resolution of issues at 95% and design completion. Effort will be made to resolve all design issues throughout the remaining design phase and verify that they have been resolved in the later design submittals and the construction documents (CDs).

The design review will also focus on the selection, evaluation, and choice of the main systems. Review the design documents against the OPR to verify that the project will meet the intent of the owner. Any choices, conclusions, or design details that deviate from the OPR will be brought to the attention of the owner and the general contractor. Additional information will be requested when documentation is insufficient to support the conclusions and choices or when required design assumptions or calculations have not been provided.

Energy efficiency is achieved by verifying the design and operation of the systems and by making the building owner aware of alternative building systems and equipment options.

Examples of building systems that will be evaluated include the following:

- Building envelope
- Building ventilation
- Lighting
- Office equipment
- HVAC equipment
- Control systems and strategies
- HVAC distribution systems
- Domestic hot-water systems
- Water use
- Occupancy schedules
- Utility rate structures

Installation Checklist Database

A checklist database will be established for all components included in the commissioned systems. The checklists will focus on providing the contractor guidance about critical requirements during installation to clearly establish the expectations of the installations.

The CxA will design these checklists to minimize the paperwork for the contractors but at the same time to cover the critical installation issues.

Construction Verification

The CxA will facilitate monthly on-site construction meetings to ensure all design, construction, and building owner representatives understand the process, the desired outcomes, and the roles/responsibilities of the various team members. The CxA will focus on training and on the Cx process during construction while at these site visits. During the construction review, the CxA will focus on verifying that the Cx process is being followed by statistical sampling and verifying that the construction checklists are completed and submitted as required. The CxA will also verify that the record drawings are on site and are being updated with any deviations in installations compared to the construction drawings. In addition, the construction progress will be evaluated against the established OPR. The CxA will verify that the Cx process is proceeding as intended during the construction phase and will review the site visit reports and Cx meeting minutes. The CxA will notify the building owner and general contractor if the Cx process is not progressing as intended by identifying and resolving issues. The day-to-day follow-up will be the responsibility of the general contractor and the subcontractors.

Review Submittals

The CxA will review the submittals concurrently with the architect and engineers. Any observed deviations from the OPR will be noted and submitted to the architect and engineers to be evaluated and submitted with their comments back to the contractor. The architect and engineers' submittal review process will also be evaluated. A selection of the architect and engineers' submittals responses will be reviewed to verify that any deviations from the design documents are properly addressed. The CxA must understand the general contractor's project delivery process and its impact on the submittal review step.

Training

The training agenda format will be submitted by the engineers to the general contractor and owner to schedule the required training sessions. The CxA will review this training agenda and attend a key training session. Each training session will be evaluated after completion of the training. Any deviations from the expected competence level of the operation and maintenance (O&M) staff will be discussed with the owner and contractors, and the remaining training agendas will be revised to accommodate any lacking knowledge.

Systems Performance

The systems performance tests will be completed as soon as all submittals for the systems manual have been received and all installation checklists have been completed. These systems performance tests will focus on the installed systems' capabilities to meet the design intent. The CxA will document the procedures required for these tests and submit these test procedures to general contractor for the project team's and general contractor's review. The subcontractors are responsible for ensuring that all systems can meet the specified requirements and for demonstrating that the systems are able to perform all procedures successfully. The CxA will witness a representative number of systems performance tests to verify that all systems work as intended. If any of the systems performance tests are unsatisfactory, these systems and a representative number of other similar systems will be required to be retested at the contractors' expense.

Review Systems Manual

The general contractor will generate the systems manual based on the subcontractor submittals for the installed equipment and the test and start-up results. The CxA will review this systems manual and provide any comments to general contractor.

Commissioning Report

The Cx report will summarize the results of Cx activities for the project. This Cx report will essentially be the Commissioning Plan with all the results of the Cx activities. The initial Cx report will be submitted two weeks after substantial completion, and the final report will be submitted one year after substantial completion. This is the responsibility of the CxA.

Operation and Warranty Review

The operation and warranty review will be completed at ten months after completion. The review will focus on the experiences of the O&M staff with the building operation and evaluate the systems performance and operation relative to the OPR. Any deviations from the original operational intent and any component failures will be noted and addressed with the owner's representative. A report will be issued to the owner with suggested actions to take.

Table C-1 Sample Commissioning Scope Matrix—Responsibilities and Schedule

Responsibility						Project Phase				Commissioning Task
Architect	Engineer	General Contractor / Construction Manager	Subcontractors	CxA	Owner	Predesign	Design	Construction	Construction	
						X				**Designate CxA (qualifications apply)**
										Provide name, firm, and experience information for the CxA
						X				**Develop the OPR; include:**
										Primary purpose, program, and use of proposed project
										Project history
										Program needs, future expansion, flexibility, quality of materials, and construction and operational cost goals
										Environmental and sustainability goals
										Energy efficiency goals
										Indoor environmental quality requirements
										Equipment and system expectations
										Building occupant and O&M personnel requirements
						X				**Develop a Commissioning Plan**
										Cx program overview
										Goals and objectives
										General project information
										Systems to be commissioned
										Cx team
										Team members, roles, and responsibilities
										Communication protocol, coordination, meetings, and management
										Description of Cx process
						X				**Implement a Commissioning Plan**
										Document the OPR
										Prepare the BoD
										Document the Cx review process
										Develop systems functional test procedures
										Review contractor submittals
										Verify systems performance
										Report deficiencies and resolution processes
										Develop the systems manual
										Verify the training of operations
										Accept the building systems at substantial completion
										Review building operation after final acceptance

Table C-1 Sample Commissioning Scope Matrix—Responsibilities and Schedule *(Continued)*

Architect	Engineer	General Contractor / Construction Manager	Subcontractors	CxA	Owner	Predesign	Design	Construction	Construction	Commissioning Task
							X			**BoD**
										Include narrative of systems to be commissioned
										Document design assumptions
										Reference applicable standards and codes
							X			**Cx requirements in CDs (include in specifications)**
										Specify Cx team involvement
										Specify contractors' responsibilities
										Specify submittals and submittal review procedures for Cx process/systems
										Specify O&M documentation requirements
										Specify meetings documentation process and responsibilities
										Specify construction verification procedures and responsibilities
										Specify start-up plan development and implementation
										Specify responsibilities and scope for functional performance testing
										Specify criteria for acceptance and closeout
										Specify rigor and requirements for training
										Specify scope for warranty review site visit
							X			**Conduct Cx Design Review**
										Review and update OPR for clarity, completeness, and adequacy
										Review BoD for all issues identified in OPR
										Review design documents for coordination
										Review design documents for compliance with OPR and BoD
										If multiple reviews are performed, check compliance with previous review comments
								X		**Review of Contractor Submittals**
										Review all product submittals to make sure they meet BoD, OPR, and O&M requirements
										Evaluate submittals for facilitating performance testing
										Review all contractor submittals for compliance with design intent and CDs

Table C-1 Sample Commissioning Scope Matrix—Responsibilities and Schedule *(Continued)*

Responsibility						Project Phase				Commissioning Task
Architect	Engineer	General Contractor / Construction Manager	Subcontractors	CxA	Owner	Predesign	Design	Construction	Construction	
								X		**Verify Installation and Performance of the Systems to be Commissioned**
										Perform installation inspection (pre-functional checklist)
										Perform system performance testing (functional test)
										Evaluate results compared to OPR and BoD
								X		**Complete Summary Cx Report**
										Executive summary
										Document history of system deficiencies/issues
										Record system performance test results
								X		**Develop Systems Manual**
										Develop systems manual in addition to O&M manuals submitted by contractor
										Include in systems manual:
										Final version of BoD
										System single-line diagrams
										As-built sequence of operations, control drawings, and original setpoints
										Operating instructions for integrated building systems
										Recommended schedule of maintenance requirements and frequency
										Recommended schedule for retesting of commissioned systems
										Blank testing forms from original Commissioning Plan for retesting
										Recommended schedule for calibrating sensors and actuators
					X			X	X	**Project Training Requirements**
										Create project training requirements document with owner
										Participate in project training session
										Ensure O&M staff and occupants receive required training and orientation
										Create and document post-training survey
										Verify and document that training requirements are met
					X				X	**8–10 Month Warranty Walkthrough**
										Perform a warranty systems review within ten months after substantial completion
										Resolve any issues found
										Create a plan for resolution of outstanding Cx-related issues

Appendix D— Early-Phase Energy Balancing Calculations

One of the key changes from the 30% to the 50% Advanced Energy Design Guides (AEDGs) is more explicitly acknowledging that designers want to design—they don't want to just stick to prescriptive measures that might constrain their creativity or the application of the design on their particular site. The authors assume that designers pick up 50% AEDGs because they want to achieve at least a 50% savings as compared to ASHRAE/IESNA Standard 90.1-2004 (ASHRAE 2004) but want some help doing it their own way (within reason). This appendix presents an overview of a method for accounting for reasonable deviations from the 50% AEDG "prescriptive" solution described in the recommendation tables in Chapter 4 while still holding to the absolute energy use intensity (EUI) value that provides for an equivalent 50% solution. This method can be used to approximate energy savings during the schematic design phase in order to inform key design decisions.

PERIMETER ZONE OPTIMIZATION METHOD

The perimeter zone optimization method is an iterative method that uses traditional whole-building energy modeling tools in a very limited way by examining only a small 10–15 ft wide piece of perimeter zone and running a series of permutations on glazing, façade, and shading configurations. This effectively is a sensitivity analysis approach that concentrates on the building envelope permutations and their potential impacts on incoming heat gain arising from the outdoors. The benefits of this approach are that there is total freedom for the façade designer, there is a consistency in the tools used between early phases and the whole-building energy modeling at later phases, and the method can be applied to any occupancy type. The drawback of this approach is that it is primarily focused on the façade, which, while important, represents only a fraction of the whole-building energy savings available. It is recommended that the perimeter zone optimization method be applied as a precursor to whole-building energy modeling to lock down the aesthetics versus energy-use questions related to the façade, as this method offers a faster turnaround time when exploring design iterations.

This method easily and efficiently provides early performance modeling by evaluating the combined impact of building orientation and envelope alternatives on perimeter-zone performance and associated heating, ventilating, and air-conditioning (HVAC) and lighting design choices and, therefore, building energy efficiency. This methodology can be employed using any number of energy modeling software currently available and is an effective early design tool/method to engage architects and owners in proactive and fruitful design discussions about siting, building configuration, and building envelope that include building energy efficiency as an equal decision influencer.

It is important to remember that there are a number of code, planning, sociological, programming, and other constraints that the design team must consider alongside options to optimize site, orientation, building configuration, and envelope energy-efficiency desires. And there are invariably times when one or the other of these design issues will be considered sacred and unable to be modified to suit the optimal solution from an energy-efficiency point of view (i.e., city-center compressed sites and land values may dictate an orientation and a building configuration that are not ideal for easy or cost-effective energy-efficient envelope design). Thus, performance/energy modeling in the early design stages must be about providing information—rather than a (singular) solution—that can inform and instigate knowledgeable decisions.

The following are a few essential key points to the perimeter zone optimization method itself.

- When modeling and discussing efficiency with the design team, it is not just about *energy* efficiency but rather must be about a balance of efficiency between any number of different *performance* indicators specific to the particular system in question. (For example, for the building envelope, *energy* efficiency needs to be balanced with thermal comfort, visual comfort, etc., to achieve *performance* efficiency. This balance represents a more complete and optimized set of interrelated efficiencies for each system, of which energy efficiency is a subset).
- The modeling analysis must include multiple issues that illustrate impacts on adjacent systems wherever possible. As an example, window shading options for different orientations may inform siting alternates and their ability to work with different HVAC options. For instance, if solar loads can be controlled, low-energy-use systems such as radiant cooling or natural-ventilation-induced cooling might be used. While it is exactly this type of impact information that facilitates the design team's decision-making process, this invariably means essentially a matrix of performance analysis options, a large number of performance analysis runs, and a great deal of data to keep organized.
- Particularly important during the initial design phases is the weighing of different options/alternatives/choices. Thus, to spark effective discussion, early design performance modeling must be comparative as well as attempt to include all of the relevant performance indicators in the analysis. This again invariably means many performance analysis options and analysis runs as well as much data to organize.
- The performance modeling/analysis results will be represented to owners, architects, and other high-level decision makers, many of whom likely do not often think in definitive numerical values (a language inherent to the engineer). Thus, again, to spark effective discussion, performance analysis results must be portrayed in a manner that the audience will understand or can quickly/easily be educated to understand; charts, graphs, graphics, and the comparative nature of the analysis all lend themselves to this and facilitate easy understanding in these types of forms. This invariably means plenty of post-processing of performance analysis results (typically numeric in their base form).

To offset the large number of analysis models and runs and the amount of data post-processing, this methodology requires that a very simple one-room model be developed and used (with each of the variables of orientation, façade treatments, HVAC options, and lighting control alternatives modified one at a time, in that same one-room model). Because there is as yet no actual building nor the details of system component choices, this simplification saves an enormous amount of model development time.

Nonetheless, experience has shown that however carefully one attempts to limit the number of alternatives to each system, the need for information involves multiple performance parameters, results, and models, each with a variety of components that must be analyzed separately. Thus, this process invariably requires a considerable amount of modeling analysis time. Additionally, the current linear design process structure doesn't typically include the technical staff needed to provide this type of information, nor is such staff typically present to be involved in representing, explaining, and discussing the results and ramifications in early design team deliberations.

The results of this method of analysis are the types of information and discussion needed for the design team to include performance efficiency (including energy efficiency) as an equal parameter in the early design decision process. These types of information and discussion are absolutely essential if an integrated approach to high-efficiency systems and buildings is a project goal. Investing in robust technical analysis early in the process is a low-cost but extremely valuable way of putting the project on the right track from an efficiency point of view.

However, to gain wider adoption of both the process (including technical staff in early design discussions) and the information (which requires performance modeling in early design phases) presented above, some key design structures and perceptions will likely need to be addressed:

- The allocation of time and money to support the effort to develop, represent, and discuss interrelated technical information in the early design phase is essential to the ability of a project team to include the effort and information so fundamental to efficient design decision making in the design of highly energy-efficient buildings and projects.

- An agreement that spending the time to include this information to enlighten design decisions early in the design process is by far "cheaper" than spending the time later to include this information to modify design decisions in subsequent design phases. The later this elemental information is included, the more potential it has to require additional time to unravel, with the resulting redesign of all the systems linked to a seemingly minor design change. (A good example is removing, increasing, or fundamentally changing the nature of external shading, which will impact not only the structural design efficiency and the cost of the glazing support members but also the redesign of associated HVAC systems, lighting systems and controls, etc.)

- There may be the perception that including the proposed discussions early in design will impose pressure to commit early, thereby limiting a designer's ability to make thoughtful, considered design decisions. However, including technically knowledgeable individuals and detailed performance information and having this information inform design decisions does happen at some point in the design process, it's just a question of when. Since many design decisions that influence integrated system performance (including energy-efficiency performance) happen early in the design process, having the information available at that time makes a lot of sense. Time can still be taken to make thoughtful, considered design decisions informed by performance analysis consequences as a part of the primary decision making rather than retroactively.

- The change in process will inevitably mean that there are "more cooks in the kitchen" during the early design phase, which can mean more project management burden imposed on the design-team leaders (the project architect, design architect, and architect project manager); however, coordination and control is needed whether it is in the early or the later stages. If appropriate time and funding can be reallocated to the early design phase, the benefits to the design (a highly energy-efficient building with less impact on relevant design aesthetics) can far outweigh this concern.

Figures D-1 and D-2 represent partial examples of typical schematic design comparative envelope and façade performance analyses (including potential energy-efficiency impacts) to illustrate performance modeling information that proved to be useful for generating discussions and informing early design decisions.

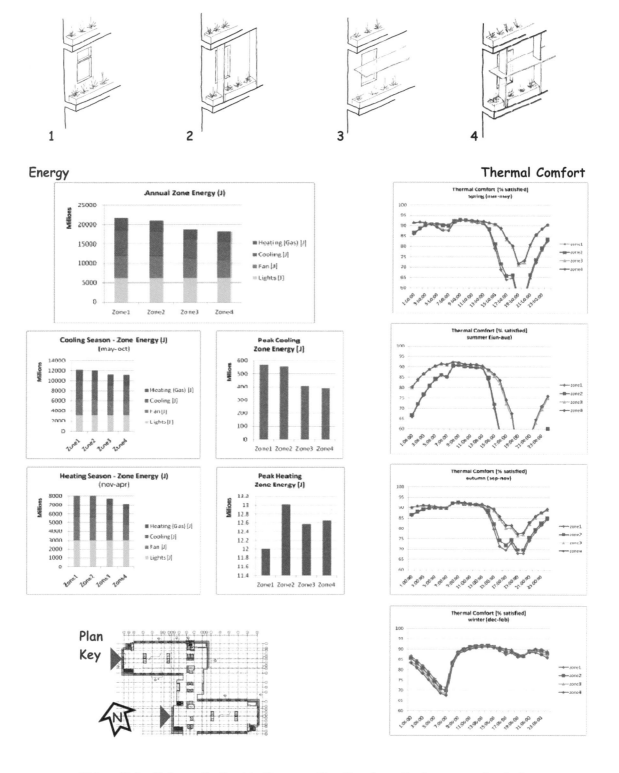

Figure D-1   Schematic Design Comparative Envelope Performance Analysis

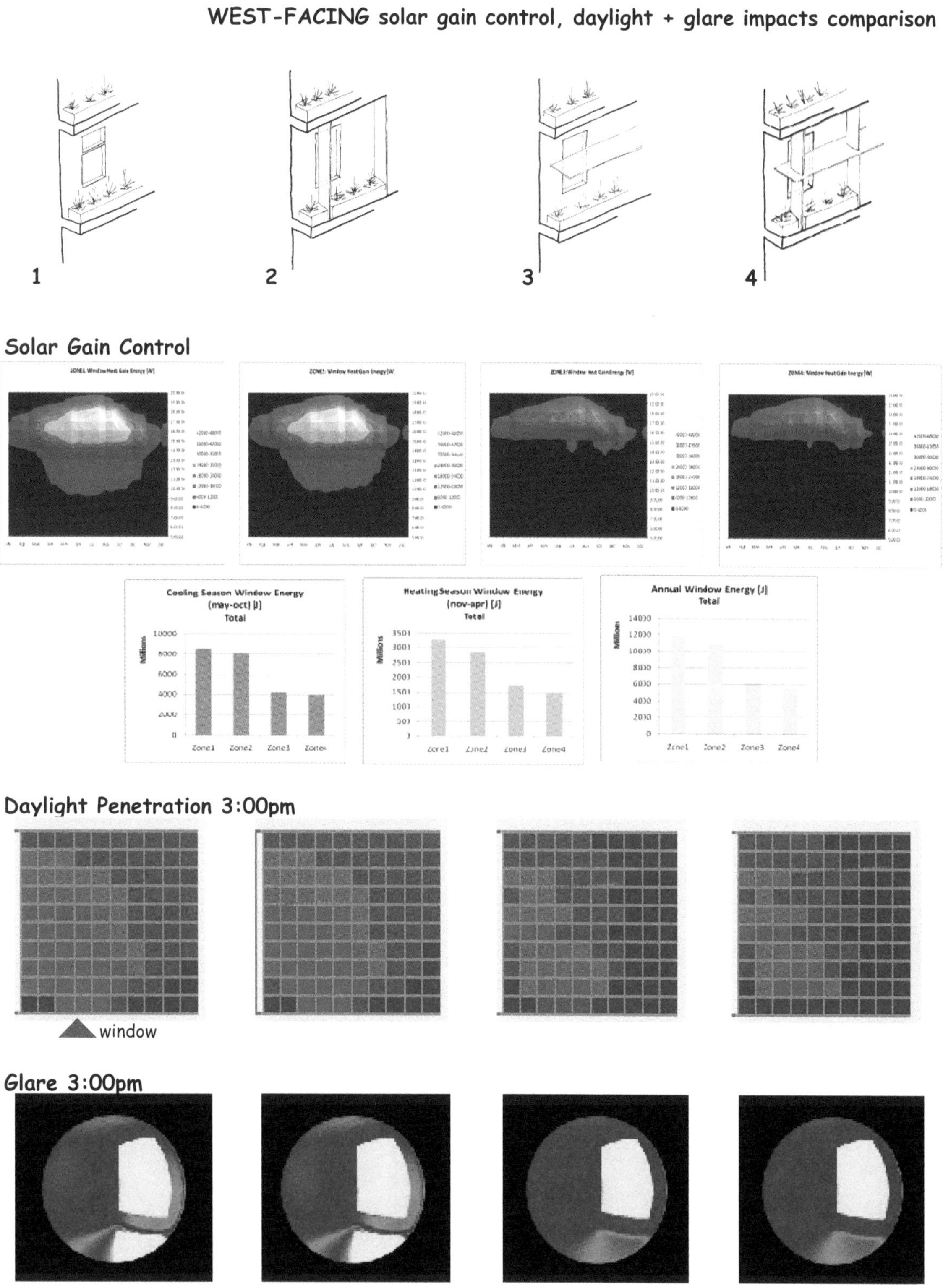

WEST-FACING solar gain control, daylight + glare impacts comparison

1   2   3   4

Solar Gain Control

Daylight Penetration 3:00pm

window

Glare 3:00pm

Figure D-2   Schematic Design Comparative Façade Performance Analysis

REFERENCE

ASHRAE. 2004. ANSI/ASHRAE/IESNA Standard 90.1-2004, *Energy Standard for Buildings Except Low-Rise Residential Buildings*. Atlanta: American Society of Heating, Refrigerating and Air-Conditioning.